永恒与变迁

（挪威）哈罗德·霍耶姆
Harald Høyem

挪威建筑师镜头下的中国

（汉英对照版）
（Chinese-English）

1985-2019

翻译 | 许东明
校译 | 王韬、董卫

PERMA-NENCE AND CHANGE
China through the Lenses of a Norwegian Architect

1985-2019

Translator | Xu Dongming
Proofreaders | Wang Tao, Dong Wei

朝华出版社
BLOSSOM PRESS

目录

CONTENTS

序言一

改革开放以后，随着社会经济的快速变革，中国逐渐吸引了越来越多外国人前来旅游观光，近距离体验和观察各地社会变化和城乡发展的方方面面。据统计，1978年入境旅游人数只有180万，而到1988年就超过了3000万。哈罗德·霍耶姆（Harald Høyem）教授就是在这个时期来到中国，与他的一些同事和学生们开始了中国城乡的研究。尤其难能可贵的是，他的这种持续性的观察、研究与思考居然坚持了30余年，真正目睹了中国翻天覆地巨变中的许多片段和细节。这些积累性的片段和细节如同历史长河中的一条涓涓细流，不仅为我们认知这段近在咫尺的历史提供了多样化的视角，同时也为其增添了一抹具有人文情怀的异域色彩。

我与哈罗德教授相识于1989年，那时我在西安交通大学建工系建筑学教研室任教。一天，学校外事部门发来通知，说要接待两位挪威特隆赫姆大学建筑系的客人。几天后，我和同事去当时的西安西郊机场，顺利接到他与夫人玛丽娅（Marie Louise Anker），还有他们不到两岁的女儿莉迪亚（Lydia Høyem Anker）。那时，交大招待所的条件不是很好，他们真切体验到了西安的一场严冬，也自此开始了我们延续至今的友谊与合作。

哈罗德教授来访的目的很明确，希望能够找到一处有文化内涵的旧居住区，一同开展改造设计研究。于是我们一同走访了城内外一些老旧住区，最终双方都认为鼓楼回民区是一个比较理想的研究对象。该社区位于老城中心，规模适中、历史悠久、人口密集、交通便利，且发展方式明显不同于周边街区，有着浓郁的中国伊斯兰文化特征。更为重要的是，这里有着相对独立的面向全市、连接西北和西南的民族市场体系，其背后则是一个更加庞大复杂的区域性回民社会经济文化网络。

PREFACE I

Since China's reform and opening up in 1978, with the rapid social and economic changes, more and more foreigners have been attracted to visiting China to experience and observe at close range all aspects of its social changes as well as its urban and rural development. The number of foreign tourists to China was estimated to be only 1.8 million in 1978, but exceeded 30 million in 1988. During this period, Professor Harald Høyem came to China and, together with some of his colleagues and students, began his research on urban and rural areas in the country. What is particularly remarkable is that he has made continuous observations, done the research, and reflected for more than 30 years, thus witnessing many moments and details of the dramatic changes that occurred in China. These accumulated fragments and details are like a trickle in the river of history, not only providing us with diverse perspectives on this recent history but also adding a touch of humanity and exoticism to it.

I met Harald in 1989 when I was teaching at the Department of Architecture and Civil Engineering of Xi'an Jiaotong University (XJU). One day I got a message from the international office of XJU that I would receive two guests from the architectural department of a university in Trondheim, Norway. A few days later, my colleague and I went to the Xiguan Airport in the west suburb of Xi'an to pick up Harald, his wife Marie Louise Anker, and their daughter Lydia Høyem Anker, who was less than two years old. At that time, the conditions of the guest house at XJU were not very good, so they experienced a really harsh winter in Xi'an. The friendship and collaboration we started then continues to this day.

此后的几年里，哈罗德教授多次与同事和学生一起来到西安，与西安交大建筑学专业的师生们对鼓楼回民区开展测绘、调查、访谈与研究，一同体验并思考这座古城的文脉、特色、问题与未来。在合作研究过程中，给我印象最深也是受益最大的一点，就是哈罗德教授与一同来西安的挪威老师们实际上形成了一个跨学科团队，其中包括城市规划、建筑学、人类学、建筑技术等不同方面的专家，他们大多有在亚洲不同国家开展城市与建筑研究的经验，善于理解不同社会与文化背景下物质环境的变迁与发展。根据鼓楼回民区的历史文化特征，我们将区内以十座清真寺为中心形成的十个传统"教坊"作为"社会—文化空间单元"分别进行走访调研。在这个过程中，我们与不少社区居民成了多年的朋友，其中包括化觉巷清真大寺的马良骥阿訇。马阿訇长期主持大寺宗教事务并兼任中国伊斯兰教协会副会长和陕西省伊斯兰教协会会长。尽管公务繁忙，但他对中挪联合调查组在回民区的工作非常支持，不仅提供了许多重要的帮助，还时常请我们去大寺访问，

咨询我们对大寺保护修复工作的意见和建议。在他的安排下，西安交大的师生对大寺及其他一些民居进行了详细测绘，系统性地学习了传统清真寺建筑的空间组织方式和相关宗教礼仪。连续数年的跟踪调查使我们深刻体会到民族自治政策及自发性的市场经济对整个回民区更新改造所产生的持续性影响，这种系统性的自治、自发与自建模式的旧区改造不仅在西安老城里是独一无二的，在当时全国的名城中也十分罕见。

通过与哈罗德教授及其挪威团队的合作，我逐渐认识到大规模更新与改造将是今后数十年中国城市所面临的重大课题，也萌生了出国深造、学习西方经验的念头。恰好这时，哈罗德教授也希望将在西安的研究发展为一个长期课题，需要稳定的合作方，于是双方一拍即合。我于1991年秋在特隆赫姆的挪威理工学院建筑系（现挪威科技大学建筑与设计学院）正式注册，成为哈罗德教授的第一位博士生。实际上此前的1990年初，我就已应邀前去该校访问

Professor Harald came to visit China with a clear purpose, hoping to find an old residential area with cultural connotations and carry out a renovation design and research project there. We visited some old residential areas inside and outside the old city, and finally we both agreed that the Drum Tower Muslim District (DTMD) was an ideal research object. Located in the center of the old city in Xi'an, with a moderate scale, this district has a long history, dense population, and convenient transportation, where the development mode was distinctly different from the surrounding areas, having a strong identity recognition of Chinese Islamic culture. More importantly, it has a relatively independent ethnic market system, facing the city and connecting the northwest and southwest of China, behind which there is a larger and more complex regional socio-economic and cultural network of the Hui ethnic group.

In the following years, Harald made several trips to Xi'an with his colleagues and students. Together with the teachers and students at the Architecture Department of XJU, we have conducted mapping, surveys, interviews, and research on the Muslim quarter of DTMD, experiencing and pondering on the context, characteristics, problems, and future of this ancient town. What impressed and benefited me most during this collaboration is that the interdisciplinary research team consisted of Harald and his Norwegian colleagues, including experts in urban planning, architecture, anthropology, building technology, and so on. Most of them, who have experienced in urban and architectural research in different Asian countries, are skilled at understanding the changes and development of the physical environment in different cultural and social contexts. Based on the historical and cultural features of the Muslim quarter of the Drum

了一次，并在他家里住了近一个月。那次恰逢他50岁生日，生日当天来了不少亲朋好友，让我着实体验了一回挪威家庭生日聚会的欢乐氛围。访问期间，我们多次讨论了西安课题的研究工作和未来学习的相关事宜。同时，哈罗德教授还介绍我认识了尽可能多的老师，从而了解到在这个国家居然有那么多人在亚洲和非洲国家开展研究工作，感受到那个时代一些挪威学者为发展中国家城乡进步做贡献的职业精神。

1991—1995年读博期间，我深刻地体会到挪威学者及年轻人对环境、资源问题的关注。甚至在建筑设计课教学过程中，师生们也用很多时间来探讨环境问题以及人类社会的未来，思考建筑设计应采取的对策。当时正值挪威女首相布伦特兰（Gro Harlem Brundtland）执政时期，她不遗余力地在全国乃至全世界推动节能环保理念。她曾任联合国环境和发展委员会主席，主持起草并于1987年向联大提交了那份著名的《我们共同的未来》（Our Common Future）工作报告。正是这份报告首次提出了"可持续发展"这一对世界产生重大影响的新思想。在这种背景下，学校里的一些教授也开设了相关课程，强调节能、环保与城乡可持续发展的重要性。因此，以西安鼓楼回民区为例探讨旧城区居民自建式的保护改造与再发展，也就自然成为哈罗德教授和我共同确定的博士论文选题。我认为，在挪威学习到的城乡可持续发展思想、全球化过程中的国际学术合作观、多学科研究模式，以及宽松平等的学术讨论方式等都对我后来的工作和研究产生了十分重要的影响。

从挪威科技大学毕业后的这些年里，我一直与哈罗德教授及其同事保持合作关系。关于西安鼓楼回民区的研究也在他的积极推动下持续深化，中方合作单位则从西安交通大学转为我的母校——西安建筑科技大学。令人欣喜的是，经过西安建筑科技大学和挪威科技大学师生多年的共同努力，当年西安交通大学的同学也曾测绘并研究过的化觉巷125号民

Tower District, we divided it into ten traditional religious quarters centered on ten mosques as "social and cultural space units" to conduct the field investigations. In the process, we became friends with many community residents over the years, including Imam Ma Liangji of the Great Mosque at Huajue Lane. Imam Ma has long presided over the religious affairs of the Mosque, and also served as the vice president of the Islamic Association of China and the president of Shaanxi Islamic Association. Despite the strains of office, he provided great support to the work of the Sino-Norwegian joint research team in the Muslim quarter. He not only provided us with much important assistance but also invited us to the Great Mosque from time to time to consult with him on the opinions and suggestions on the conservation and restoration works of the Great Mosque. Under his arrangement, the teachers and students of XJU conducted detailed surveys and mapping of the Great Mosque and some other residential buildings, so that we were able to systematically study the spatial organization of traditional mosques and related religious rituals. Additionally, the follow-up fieldwork over the years made us fully realize the continuous influence of the national autonomy policy and the spontaneous market economy on the renewal of the Muslim quarter. The systematic renovation of old blocks, which has been conducted based on autonomy, spontaneity, and self-construction, was not only unique in Xi'an but also very rare among other renowned historical cities in China at that time.

Through the collaboration with Professor Harald and his Norwegian team, I gradually realized that large-scale urban renewal would be a major issue for Chinese cities in the next few decades, and I started to think about studying abroad and gaining some experience

居保护项目获得了2002年联合国教科文组织亚太地区遗产保护奖。作为该奖的评委之一，我为此感到由衷的高兴。

与此同时，哈罗德教授的同事们还在拉萨以及尼泊尔加德满都、印度加尔各答等城市开展研究。其中克努德·拉森（Knud Larsen）教授编纂的《拉萨历史城市地图集：传统西藏建筑与城市景观》是他多年研究的成果。为此，西藏大学与挪威科技大学分别于2003年和2005年在拉萨组织了两次关于拉萨老城保护研究的国际研讨会，我均应邀参加。会上与哈罗德教授、克努德教授、汉斯（Hans Bjønness）教授等多位挪威朋友再次相逢。2010年上海世博会前后，哈罗德教授与他的同事们多次来东南大学建筑学院交流，就江南历史城市保护更新开展联合教学。

从2012年开始，我们有机会对平遥地区的60余座传统村落开展调查研究。当哈罗德教授提出希望在这方面开展一些合作时，我推荐了距离高铁站比较近的杜村、侯冀和桥头等乡村，最终双方确定以侯冀村作为研究对象，研究题目聚焦于传统民居的生态化利用。2014—2015年期间，我们对所选定的民居群落在详细测绘的基础上进行了长达一年的实时环境监测，掌握了四季全时段的室内外气象及环境数据，据此提出了基于物理环境优化的民居改造设计思路。

对我而言，读哈罗德教授的书就如同回顾一段亲身经历的历史，许多看似直白的文字和普普通通的照片都能够激发出一些鲜活的记忆，这一切仿佛就像发生在昨天一样。他所提到的许多人物我也十分熟悉，看到他们的名字，我的脑海里就会闪现出一个个生动的形象。因此，在写这篇序言时，难免会沿着一种回忆式的思绪，为他书里的某些章节提供一些补充性文字，聊作一种个性化的记录。

我相信从哈罗德教授的这本书里，读者可以体验到他以一位挪威建筑师的眼光、以平视的角度，从近距

in Western countries. At that time, Harald also wanted to develop his research in Xi'an into a long-term project, which needed a stable partner, so we were able to develop our relationship and work together as partners. I was officially enrolled in the Department of Architecture at the Norwegian Institute of Technology (NTH, now the NTNU, Norwegian University of Science and Technology) in Trondheim in the fall of 1991, and became the first Ph.D. student of Harald. In early 1990, I visited the NTH and Trondheim once and stayed at Harald's house for nearly a month. During my visit, which coincided with his 50th birthday, I was able to experience the joyful atmosphere of a Norwegian family birthday party with many of his friends and family members. We also discussed our research on the Xi'an project and some related matters for my future studies. At the same time, Harald introduced me to as many other professors as possible, so that I could learn that many experts in Norway were actually doing research in Asian and African countries, and making their contribution to the urban and rural progress of developing countries.

During my doctoral years from 1991 to 1995, I also deeply felt how much Norwegian scholars and young people care about environmental and energy use issues. Even in architectural design classes, teachers and students would spend a lot of time discussing environmental challenges and the future of human society, thinking about what should be done with architecture, rather than rushing students to come up with their own design proposals as soon as they started, as we usually did in China. This was the time when Norwegian Prime Minister Gro Harlem Brundtland was in power, and she spared no effort in promoting energy conservation and environmental protection

离展现出一系列日常而专业的画面和批判性的思考。这些都为我们提供了一幅以中国改革开放为宏观背景的城乡变迁动态图景。同时这本书也建构起一种历史网络，把平时远隔千山万水素不相识的人们关联起来，把西安这座古城置于全球化的框架中加以审视。这使我们有机会反复体会这30余年间城市的发展和变化，反思和检视在过去这样一段历程中中国城市发展所取得的成功和存在的问题。毫无疑问，书中所描绘出的城市历史性样貌会随着国家的快速发展而显得越发珍贵，每个人都会从中感受到中国改革开放进程所带来的巨大能量。

在此，我向哈罗德教授、他的夫人玛丽娅，以及他的同事们再次表示衷心的感谢！玛丽娅多次携女儿随同哈罗德教授到访中国，她毕业于特隆赫姆挪威理工学院建筑系（现挪威科技大学建筑与设计学院），当时在南特伦德拉格郡遗产保护部门工作，现在则是特隆赫姆市遗产保护部门的主管。1989年刚来西安时他们的女儿莉迪亚还不到两岁，而现在，她已经

是一位母亲，而且女承父业，她从父母的母校毕业后成为一位职业建筑师。我也非常希望哈罗德教授所建立起来的这种合作机制能够持续下去。在这个过程中他培养出了多位中国学生，他们在不同的工作岗位上不断做出自己的贡献，也在新的历史时期继续关心并见证中国和挪威的进步与发展。

董卫
2021年8月9日
于东南大学前工院

throughout the country and even the world. Prior to becoming prime minister, Brundtland worked as the chair of the World Commission on Environment and Development (WCED) of the United Nations, and presided over the drafting and presenting of the famous *Our Common Future* report to the General Assembly in 1987. It was this Brundtland Report that first introduced the concept of "sustainable development," a new idea that would have a significant influence on the world. In this context, some professors at the NTNU also set up relevant courses to emphasize the importance of energy conservation, environmental protection, and sustainable urban and rural development. Therefore, it was only natural that Harald and I decided the topic of my doctoral dissertation, to be taking DTMD in Xi'an as an example to explore the conservation, renovation, and redevelopment of the old urban areas in the

form of self-built housing. I believe that the ideas of sustainable urban and rural development that I learned in Norway, the approaches to international academic collaborations in the context of globalization, the multi-disciplinary research framework, and the academic discussion in a relaxed and democratic atmosphere have all had a profound impact on my subsequent works and research.

In the years after my graduation from the NTNU, I have continued collaborations with Professor Harald and his colleagues. The research on DTMD in Xi'an has also been strengthened under his active promotion, and his Chinese partner has shifted from XJU to Xi'an University of Architecture and Technology (XAUAT), which is my alma mater. It is gratifying that after years of joint efforts by the teachers and students of the XAUAT and NTNU, the conservation project of No. 125 Huajue Lane courtyard, which was also mapped and studied by the students of XJU, won an Honorable Mention from UNESCO Asia-Pacific Awards for Culture Heritage Conservation in 2002. Being one of the jury members of the award, I was sincerely happy about this result.

Meanwhile, Professor Harald's colleagues also conducted research in Lhasa as well as in Kathmandu, Nepal, and Kolkata, India. *The Historical Urban Atlas of Lhasa: Traditional Tibetan Architecture and Urban Landscape*, compiled by Professor Knud Larsen, was the result of years of his effort. To this end, Tibet University and NTNU organized two international seminars on the research of the conservation of old city in Lhasa in 2003 and 2005, respectively. I was invited to participate in both of them. At the seminars, I met again with Professor Harald, Professor Knud, Professor Hans Bjønness, and many other Norwegian friends and colleagues. Before and after the 2010 Shanghai World Expo, Professor Harald and his colleagues came to the School of Architecture at Southeast University (SEU) several times to carry out joint teaching programs on the preservation and renewal of historical towns in the south of the Yangtze River.

Since 2012, we have had the opportunity to conduct research on and fieldwork in more than 60 traditional villages in the Pingyao area. Therefore, when Professor Harald proposed some collaboration in this regard, I recommended the villages of Ducun, Houji, and Qiaotou close to high-speed railway stations. Finally, we decided to take Houji Village as the research object, and the research topic was to focus on the ecological utilization of traditional houses. From 2014 to 2015, we conducted a year-long real-time environmental monitoring of the selected vernacular dwellings based on detailed surveys and mapping, mastered the indoor and outdoor meteorological and environmental data in all seasons, and put forward the design proposal for dwelling renovation based on physical environmental optimization.

In short, for me, reading Professor Harald's book is like reviewing a history of my personal experience, and many of the seemingly straightforward words and ordinary photos summoned my vivid memories as if it happened yesterday. I have reunited with many of the people mentioned in the book, and their faces flashed before me when I saw their names. Therefore, in writing this preface, it is natural for me to provide some supplementary words for some chapters of this book as some sort of reminiscence of my experience, just to make a personalized record.

For readers of this book, I believe that they can experience a series of daily and professional images as well as critical reflections from an equal and open-minded perspective of a Norwegian architect. These provide us with a dynamic picture of urban and rural changes in the macro context of China's reform and opening up. At the same time, the book also constructs a historical network, connects people who are usually thousands of miles away from each other, and examines the ancient city of Xi'an in the framework of globalization. Through this book, we have the opportunity to recapitulate the evolution of cities over the past 30 years as well as reflect on and examine the outcomes and challenges of China's urban development. There is no doubt that the historical urban landscape depicted in the book will become increasingly valuable with China's rapid development. Everyone will feel the tremendous momentum brought by China's reform and opening up.

Once again, I would like to express my sincere appreciation to Professor Harald Høyem, his wife Dr. Marie Louise Anker, and their colleagues. Marie accompanied Harald on many of his visits to China with their daughter. She graduated and obtained her Ph.D. from the Faculty of Architecture at NTNU. At that time, she worked as the conservation officer at the Sør-Trøndelag County and is now the senior advisor of the Nidaros Cathedral Restoration Centre (NDR) in Trondheim. When they first came to Xi'an in 1989, their daughter Lydia was less than two years old. Now, she is a mother and has chosen to continue her father's profession as an architect after graduating from her parent's alma mater. I really hope that this collaboration network established by Professor Harald will continue. In the process, he has trained a number of Chinese students who have also made their own meaningful and continuous contributions in different areas, as well as shown their care for and witnessed the progress of both China and Norway in this new historical era.

Dong Wei
Southeast University, Nanjing
August 9, 2021

序言二

按照中国人的习惯，学生为导师的书写序可以被看作一种僭越。因此，收到霍耶姆教授的邀请我深感惶恐。但是换个角度去想，多年的受教、合作与日常相处，我和霍耶姆教授有了一种亦师亦父亦友的关系。作为他多年与中国相关的各种工作的参与者和旁观者，我可以提供一些有关这本书的佐证。另外，老先生不拘一格地邀请我来写，我挺身而出接受这个任务，是不是也可以看作跨文化交流产生的打破各自文化常规的结果呢？

作为一个在霍耶姆教授的故乡挪威生活工作了多年的中国人，我的个人经历可以说是此书的一个逆向佐证。书中困扰霍耶姆教授的问题也在困扰着我——作为一个外国人我该如何去理解挪威文化？有哪些关于挪威的成见在日常生活中被打破？如何从对一个文化的浅薄而笼统的认识走向和那个文化中一个具体的人的沟通和交流，从而更深刻地理解他／她的文化？

篇幅有限，这些问题留待以后再议。这里我想说的是这本书的起点——霍耶姆教授作为一位建筑学教授对文化问题始终如一的重视。在20世纪90年代初期与中方合作开展的回民区改造研究中，首先进场调研的是人类学者，这也是我生平第一次接触到人类学，这种工作方式令我耳目一新，但对其作用却不明就里。后来，当霍耶姆教授在中国和挪威开授研究生课程的时候，人类学方法成为学生进入现场工作前必须掌握的技能。此后，可以看到，人类学方法也成为他多次在中国开展研究中不断使用的工具。为什么霍耶姆教授对人类学在建筑研究中的使用如此执着？我逐渐认识到，这是因为人类学研究的对象是文化，建筑学工作的对象正是文化的载体。建筑与文化的关系是贯穿霍耶姆教授职业生涯和个人经历始终的主题。

PREFACE II

To write a preface for one's own teacher's book would be considered inappropriate in Chinese culture. So, it troubled me a bit when I received this invitation from Professor Harald Høyem (whom I will address as Harald in the following text, as I usually do in our conversations). However, with years of being instructed by, cooperating with, and daily getting along with, Harald has become not only a teacher but also a friend and sometimes a father figure to me. Thus, as a witness and participant of his decades of experiences in China, I am surely qualified to attest to the contents of this book. Besides, isn't his inviting me to write the preface and my accepting it, evidence of what is being depicted in this book – cross-cultural exchange and mutual influences?

Moreover, as a Chinese who has lived in Harald's hometown, Trondheim, for several years, I can also provide some reverse experiences to support this book. The questions which puzzled Harald also puzzled me – How to approach Norwegian culture as a foreigner What prejudices were broken when I actually lived in Norway? What has changed to the stereotyped impressions aforehand when interacting with foreign people in their own culture?

To sum up, my answers would demand another book. What I want to focus here, in the preface of this book – is Harald's persistent focus on culture. In the early 1990s when he and his team were conducting research in Xi'an Muslim district, anthropologists were the first to enter the site. Later on, anthropologists' methods became the research methods his students must apprehend in his urban renewal and heritage protection courses both in China and Norway. Eventually, anthropological methods have become recurrent tools penetrating his

自现代主义倡导建筑和城市中的功能主义设计以来，人被剥离了文化背景，建筑和城市按照生物性和功能性原则来设计。后现代以来，虽然对这种功能主义提出矫正和批评，但是更多的时候是在顺应一种新的全球性消费文化。而霍耶姆教授与众不同的地方，是对植根于一个特定场所的具体个体、家庭和社群的关注，他感兴趣的是生动、鲜活的文化，并努力去了解这种文化，为其做适宜的设计。因此，本书内容既有对一个社区或村镇的观察，也有生活在其中的个人的故事；既有对当下空间格局和日常生活的描述，也有对于历史、地理、风水等与地方性的形成紧密相关的要素的思考；更重要的是，这种跨越数十年的观察正好是中国急速变化的一段时间，书中所提到的很多事情、行为、场所对于中国人自己也都已经成为遥远的记忆，于是此书也恰好成为一个快速发展时代的记录。霍耶姆教授在书中也试图去捕捉和理解这种快速发展下的文化变迁，分辨固定不变的部分和变动不居的成分去指导具体的建筑与城市设计。

研究建筑自文化开始，这正是西安交通大学"饮水思源"精神的体现。建筑始终服务于具体的人、家庭和社群，以及他们的生活、习惯和传统，包括他们所经历的变化。通过此书，霍耶姆教授生动且深刻地告诉了我们：鲜活而具体的文化始终是建筑设计的本源。

王韬
2021年6月25日
于清华大学

professional experiences in China. Why? In my understanding, the object of anthropological study is culture. Culture is embodied in human behaviors and must be carried out in certain places. These places are the products of architectural design. Therefore, the relationship between culture and architecture has become the perpetual focus of Harald's professional career and personal experiences.

With the rise of functionalism in modern architecture, the culture elements had been stripped away. At the climax of its reign, buildings and cities were designed according to biological and functional principles. Later on, although post-modernism was alleged to correct the effects of this trend, in many cases it is merely an adaptation to the global consumer culture, another round of cultural deconstruction, in economic terms. So, approaching architecture via the understanding

of local culture makes this book very special. Also, the definition of culture here is not in its general sense, but in the vivid individual, family, and community lives and their daily physical settings. Therefore, in this book, there are not only observations of a village or city, but also stories of individuals living in it; there are depictions of spatial characters and daily lives, as well as reflections on the brewing historical, geographical, fengshui, and other elements behind them. Furthermore, Harald's decades of observation coincide with the most rapidly developing period of modern China. Many things, places and behaviors are now but only found in the remote memories of the Chinese people, which makes this book an authentic record of this time of change. In this book, the readers can see through Harald's eyes, his efforts to capture and understand the culture and its transition, discerning the unchanging and the changing, eventually directing his urban and architectural studies.

Approaching architectural studies through understanding culture, this is the manifestation of the slogan of Xi'an Jiaotong University - "When drinking water, think of the source". Architecture serves people. Their lifestyles, customs, and traditions, as well as the changes they experience, are always inspirational sources.

Wang Tao
Tsinghua University, Beijing
June 25, 2021

动机

如果说过去25—30年，中国的发展速度日新月异，这一点不耸人听闻。我写此书的目的，就是想通过文字和图像，来记录自己这些年对这种变化的亲历。以我所观察到的现象，来解读说明这段时光。我想深入了解这些大变化之下人们是如何应对的。跟上发展的速度是个挑战，当下的经验可能在几年后就过时了，但回首以往，我们又看到过去的很多东西依然存在。这种从算盘到互联网贸易、古老文化和现代生活的混搭既迷人又有趣。

为什么对中国如此着迷？我很想从头讲起，但这个开头很难界定。像大多数开头一样，这个开头也有一个过去。作为一位挪威人，可能的一个起点，是当毛泽东去世后的几年，历史研究开始对他的生活和政治提出与以往不同的诠释，我的许多同胞（远远多于宣称自己是左派的人数）感到非常失望、沮丧或不快。究其原因，也许在我们这两个看似迥异的文化中，存在着一些深层次的联系？随着时间的推移，我一直在尝试更好地理解这种联系。

1985—2019年，我在中国不断旅行，给中国和欧洲的建筑专业学生授课，并在文化遗产和城乡历史住区的保护发展领域开展研究，我对中国的迷恋由此而生。试图理解文化间的异同，是贯穿本书的主导视角。显而易见，当今快速的城市化进程为日常生活提供了全新的前提，因此，观察城乡不同背景下历史住区的共同特征是一项非常有趣的工作。

对于那些生活与我们迥异的人群，我们都会有种种刻板的成见。随着时间推移，新获得的经验会使原有的成见发生微妙的变化，或逐渐淡化，或代之以新的认识。我很快意识到，每一位中国人和世界各地的人一样，都是独特的个体，因而他们也是多层次、多方面的。一开始会让我哑然失笑的事，经过

INTRODUCTION

Motivation

Pointing out that during the last 25–30 years China has developed incredibly fast would hardly be sensational. What I want to achieve, through text and pictures, is to describe my own experiences in China during these years, along with phenomena I have observed, to illustrate this period of time. I'm trying to get under the skin of the big changes and to learn how people tackle it all. It has been a challenge to keep pace with the speed of development as experiences from one year might be outdated just a few years later. As we look back, we see that much of the past is still present, which is both fascinating and interesting, this mix of old culture and modern life, from abacus to internet trading.

Why this fascination with China? It may be tempting to start at the beginning, a beginning which is difficult to define because like most beginnings this beginning too has a past: As a Norwegian, I may find a departure point in marveling at why so many of my countrymen and women - numerous compared to how few who really considered themselves to be politically active on the far left - reacted as they did years after Mao's death when historical research cast a more diverse light over his life and politics. A lot of people were much disappointed, upset, and offended. Why? Maybe there are some deep-seated similarities in the two evidently differing cultures of our countries? My attempts to understand better this emotional relationship have evolved over time.

My fascination with China took hold in the period from 1985 to 2019 through traveling, teaching Chinese and European architecture students, and carrying out research projects,

思考后会变得不那么可笑。而且，我也意识到自己的文化中也存在着类似特征，这些特征也许程度不同，也许在语境上有些差异，但究其根本殊途同归。我希望本书中收录的文字和图像能证明这点。

回想起来，我意识到了人类学家弗雷德里克·巴特（Fredrik Barth）在《他者的生活和我们自己的生活》（Andres liv – og vårt eget）中所说的一句话中蕴含的真理："一个人通过参与另一种文化更深刻地理解自己的文化。" *1

由于这本书同时使用文字和图像来剖析一个时代和人们的生活，我想对这些图像补充几点说明：

中国有大约14亿人口，因而我们很容易忽略个体，使他们消失在人群中。面对群体，你很难看到个体，但是，当一切凝固在图像中时，你可以不被所观察对象的目光对视干扰，沉浸在对他/她的观察中，看到他们种种吸引人的、有趣的活动，去诠释和想象他/她的性格、背景、梦想、成功与失意，以及从事的工作……尽管这些想象不一定符合现实，但对于视觉解读的魅力而言，这并非紧要。我们该如何避免偏见？误解又是多么容易产生？这是我们必须面对的风险。因此，尽管我选择了书中这些图像来支持我的文字，但也为读者形成自己的解读留出了充分的空间。

自珍之见

多年前，我担任一个关于建筑文化遗产保护与发展主题竞赛的评委。评审组在西安建筑科技大学工作，成员有中国人也有外国人。评审组组长是一位外国人，因为觉得我们这些外国人对中国的文化历史背景和当前社会状况均所知有限，感到非常沮丧。我们进行了简短的讨论之后得出的结论是，当外国人

*1 Barth, Fredrik. Andres liv – og vårt eget. Oslo: Universitetsforlaget. 1991.

mainly within the field of cultural heritage and the development of historic housing areas in villages and cities. Trying to see the differences and similarities between our cultures has been my guiding perspective throughout this book. It is evident that the rapid urbanization of today gives totally new premises for daily life, so it has been interesting to observe the common features found in the historic, inhabited housing areas in urban and rural contexts.

We all have stereotypes about people who are different, living under conditions deviant from what we do ourselves. As we gain new experiences over time, the original stereotypes will subtly change, or they will fade out, maybe to be replaced by new ones. I, of course, relatively soon found that Chinese people, just like people everywhere, are distinct individuals. The picture is multifaceted. Things I, in the beginning, might laugh or smile at, upon reflection, appear less comical. Not least when I realize that similar features exist in my own culture. Maybe they exist in a different shade, maybe with some differences in context, but they are still very much the same. I hope the text and pictures will demonstrate this.

In retrospect, I have realized the truth about what the anthropologist Fredrik Barth underlined in his book *The Life of the Others and Our Own*: "One learns much about one's own culture by being engaged in another culture." *1

Since this book uses both text and pictures to delve into the time and the people's lives, I'd like to add just a few comments on the pictures:

Knowing that there are around 1.4 billion Chinese citizens, it's easy to forget the

*1 Barth, Fredrik. Andres liv – og vårt eget. Oslo: Universitetsforlaget. 1991.

被邀请进入评委会，是为了从外部征求意见与想法，而不是做出我们自以为与中国人观念一致的所谓"贡献"。

这件事很好地说明了一个两难的问题。一个外国人来到中国工作并与中国同事合作，应该记住，他看到和谈及的东西，可能是一位土生土长的中国人看不到和不会谈及的。这种情况带来的影响既可能是积极的也有可能是消极的，同时也取决于如何表达。挪威人的非正式交流风格与中国人信息沟通和传达的位序层级结构相当不同。在我们结束西安鼓楼回民历史街区工作的几年后，一位曾参与该项目的退休官员告诉我，我们当时的工作方式在当地行政系统造成了一种"可怕的混乱"。但他又笑着补充道："但这对我们大有裨益。"

而这就是我们那时和此时的形象。一群金发碧眼的斯堪的纳维亚人，举止古怪，粗鲁无礼，讲着奇怪的语言。日复一日，年复一年，我学会了稍微修正自己

的方式。但是，我们总是要在对外国文化价值持开放态度和忠实于自己的文化价值之间寻求平衡（当然，这些价值也在不断变化）。试图理解是一回事，完全接受则是另一回事。有时候，要把这两者分开很难。就像有人说的那样，这就像同时使用两个不同的罗盘摸索前行。

在电影《迷失东京》（Lost in Translation）中，比尔·默里（Bill Murray）饰演的老影星来到东京拍摄一个威士忌广告。有一幕是一位激情洋溢的舞台导演对他用响亮的声音和生动的手势说了好几分钟，而当默里问翻译："他说了什么？"面无表情的翻译简洁地答道："右边，要有力度。"

这让我想起自己因为不会说中文，也不是在中国文化中长大，所以很难理解对方说的是什么、发生了什么。我们在与讲中文的人对话或采访时，总是要依赖翻译。无疑，其中会有许多误解。当然，从一个更为笼统的角度来说，有人会质疑一个人是否有

individuals and let them disappear into the crowd. It is hard to see the individual when confronted by many. It is easier when the situation is frozen in a picture, when you can study each face in the crowd, and when you are not distracted by a mutual gaze. You can then abandon yourself to the attractive and interesting activity of interpreting and imagining the personality of the individuals, their background, dreams, successes, disappointments, their line of work, and so on. Whether your speculations are in line with reality is, naturally, of interest, but not always decisive for the fascination of dealing with visual interpretations. How should we survive without our prejudices? And how easy is it to misinterpret everything? This is a risk we have to take. Thus, while the selection of pictures is my own photos intended to complement the text, I also leave space for the reader's individual interpretation.

Reservations

Some years ago, I was a jury member for a competition on the protection and development of architectural, and cultural heritage. The jury worked at Xi'an University of Architecture and Technology (XAUAT) and had both Chinese and international members. The jury leader, a non-Chinese, was frustrated because she felt that we foreigners did not have enough background knowledge of Chinese culture or on present socio-cultural conditions. A brief discussion took place and concluded that when foreigners had been asked to be on the jury, it was to get views and ideas from outside, not to have dubious contributions of what we believed would be congruent with Chinese conceptions.

This episode illustrates quite well a dilemma. When coming to China from another culture, to do work there and co-operate with

可能完全理解另一个人。而现在，我们面临的情况比平时与家人聊天要复杂得多。我的一位经验丰富的人类学同事让我明白，理解有不同的层次。如果你读了关于另一个国家的书，你会知道一点；如果你以旅游者的身份去到那里，你会从你的所见所闻中了解更多，例如和出租车司机的聊天；而当你在那里待上较长时间，和当地同事一起工作，你会有更深入的了解。但是，在理解的当下我们需要暂时忘掉是否还有比这更深层次的理解。归根结底，我们永远是外国人。当然，我们不应停止尝试，而且我们也没有这么做。

显然，语言问题有时候的确是绊脚石。下面是一个极端的例子：在一个研究项目中，我们和中国同事在一个偏远的村庄合作。我们的中国同事听不懂当地人的方言，所以我们需要一位翻译将当地土话翻译成普通话，然后我们的中国同事再把普通话翻译成英语；而当我们想做出回应的时候，需要把这个顺序反转，这确实耗时耗力。此外，英语甚至不是

我们这些外国人中任何一个人的母语，这无异于雪上加霜。但也可能并非如此？也许这反而是一个优势？也许正是因为我们必须在这样一个基本的语言层面进行交流，较少去依赖文字表面意义，从而迫使我们对字里行间的内容更加开放？也许它迫使我们对人与人之间其他层面的交流更加敏感？

我想读者现在会提出一个非常关键的问题：你们是否完全明白自己在那里做什么？我坚持认为，在一定程度上，是的，我们明白。而且随着时间的流逝，我们渐渐理解得越来越多。知识和理解有很多种。凡事积于跬步，起于寸土片石。在实践中学习，有时也从错误中学习，我们会建立一种默契的知识。当然，你也可以一直质疑这种知识的有效性如何，我希望读者可以在阅读中自己去判断。

Chinese colleagues, one should keep in mind that a foreigner may see and say what is not possible or likely for a native Chinese to see nor say. This may have positive or negative effects, also depending on how things are expressed. The Norwegian informal style is rather different from the Chinese style of hierarchic order of communication and distribution of information. Some years after we had finished our work in the Drum Tower District, a retired high-level official, who was involved in this project, told me that our way of working had caused kind of frightening chaos in the municipal systems. "It was very good for us," he added with a laugh.

So that has been our role now and then: A horde of fair-haired, blue-eyed Scandinavians, with odd behavior, saying strange things in an impolite and rude way. Year by year I learned to slightly modify our style. But we always had to look for a balance between an open mind for the values in this – for us – foreign culture and at the same time dedication to the values of our own culture, (values which are, of course, also changing). Trying to understand is one thing, and to accept is quite another. Sometimes it is difficult to keep the two apart. One could say it's like muddling through with two compasses.

In the movie *Lost in Translation*, Bill Murray plays an elderly movie star coming to Tokyo to film an advertisement for a whiskey brand. In one scene he receives instructions from a very enthusiastic and expressive stage director who talks for several minutes in a loud voice and with vivid gestures. Murray asks an interpreter "What did he say?" With a poker face, the interpreter replies: "Right side, and with intensity."

This reminds us of our own difficulties understanding what was being said and what was going on, not speaking Chinese, nor having been brought up in the Chinese culture. We

关于本书

有时候，一个人会在与自己的祖国截然不同的环境中度过人生的大段时间。在这种情况下，许多紧迫的问题需要求得答案。我们如何与语言不通的人沟通？我们如何看待一种明显不同于我们自己的文化？在一个以不可思议的速度变革的国家，我们如何理解时间的概念？还有就是我在本书中试图解答的主题：我们在建筑领域活动中如何理解语境？——我基于自己在中国30多年的经历，尝试就这些问题进行一些思考。

本书将以各种不同的方式关注"永恒和变迁"这一主题，以及文化之间的异同。全书内容围绕两个轴线展开：其一是我的主要职业趣旨——居住和文化遗产；其二是在这段时期，我对中国发展的个人感受和专业经验。还有两个方面的内容本来可以纳入此书之中。一个是这35年中我的中国同事和朋友们各自开展的高质量研究，但是其内容太过庞杂难以处理，

可能需要另外一整本书去专门论述。另外一方面的内容是我有意回避的，那就是悠久而丰富的中国建筑历史本身。在本书中，我选择将内容专注于我们这个时代日常生活中的建筑和人。

第一章讲述了我与中国的初次接触。1985年，我在中国待了七周，研究中国地理环境下的风土建筑。这次旅行的一个结果就是我对文化差异的思考发端。从一个遥远的国度如何慢慢去接近中国。

第二章介绍了1985年我所看到的西安，以及我与中国同事的初次接触，这种接触后来发展成为长达34年的持久联系。这一章既有与中国同事和普通人接触的详细观察和介绍，还有1985年访问时西安及周边地区的情况描述。

第三章讲述了1989年我在西安的首次长期逗留和我在西安交通大学校园的生活经历，以及四年以后再访问中国，对于西安交通大学住宅区和鼓楼回民区

were always dependent on talking through a translator when having conversations or interviews with Chinese-speaking people. No doubt there were numerous misunderstandings. Of course, one could question whether it is at all possible to fully understand another person – to put it in a very general perspective. And now we were faced with much more complicated situations than we usually were when chatting with people at home. Our experienced anthropologist colleague made clear to me that there are different levels of understanding: If you read about another country, you will know a little. If you go there as a tourist, you will understand better, from what you see and from listening, for instance, to the taxi driver. When you stay there for an extended time, working with local colleagues, you get an even deeper understanding. But forget about trying to understand on a deeper level than that. In the end, we will always

be foreigners. Still, of course, one shouldn't stop trying. And we didn't.

Evidently, language problems can be a real stumble. Here is an extreme example from a research project where we collaborated with Chinese colleagues in a remote village: Local people there had a dialect our colleagues did not understand, so we needed a local interpreter to translate from the local dialect to standard Mandarin. Our Chinese colleague then translated from Mandarin to English. And then the sequence was turned around when we wanted to say something in return. It was complicated and time-demanding, indeed. And, furthermore, English was not even the mother tongue of any of the participants in our activities in China, which did not make things easier. Or maybe not? Maybe this was an advantage? Maybe the fact that we had to communicate on such a basic language level, less dependent on the ar-

的居住状况的了解，成了此后持续关注中国快速发展的良好起点。这一章介绍了西安旧城中心鼓楼回民历史街区工作的开端，这项工作后来持续了大约20年。

第四章介绍了1989年西安鼓楼回民区的情况，以及那里发生的变迁。这一章还介绍了挪威科技大学与中国西安交通大学在1991年的第一次硕士生联合教学。学生们就西安鼓楼回民区的历史居住建筑进行研究，调查了彼时人口密集居住地块的问题和潜力。1989年对西安鼓楼回民区居住条件的首次研究，以及后来与中国同事合作向欧洲建筑专业学生呈现问题的复杂性，加深了对那里情况的了解。

第五章深入探讨了面对各类文化差异时专业人士的角色和挑战。在这个时期，中挪签订了有关西安鼓楼回民区部分区域发展的协议，内容涉及方方面面。从1997年到2003年，该项目持续了六年时间。本章讲述了在一个国家级合作协议框架下，挪威科技

大学团队在中国工作必须渐次跨越的几道门槛：亚洲—中国—西安—回民区。

第六章是关于1997—2003年期间对西安鼓楼回民区日常生活中的居住文化遗产的研究，深入到该片区的具体现实（这种现实在其他地方也很有代表性）。我在此讨论了关于保护和维修的国际宪章，并说明了这些文本与西安鼓楼回民区项目的关系。在国际上对于风土民居保护问题不断增长的关注下，中国的文化遗产部门也开始注意这个问题。关于保护所遵循的原则问题在西安鼓楼回民区遇到了挑战。

第七章讨论的是户外公共空间的活动，这些活动似乎是长期稳定的，很少发生变化。如打麻将、打牌、下棋以及其他公园活动，都以图像说明。在户外活动中，永恒与变迁的关系可以看得很明显。许多有着悠久传统的活动没有改变，而另一些则经过了现代化改造或采取了其他的新形式。

ticulated meaning of the words, forced us to be more open to what was said between the lines. Maybe it forced us to be more sensitive to some other level of inter-human communication?

Thus, I suppose the reader now will ask a very crucial and critical question: Did you at all understand what you were doing there? And I will maintain that yes, to a certain degree, we did. And gradually, as time passed, we understood even more. There are many kinds of knowledge and understanding: Moving step by step, building stone by stone, learning by doing, and sometimes learning from mistakes, we build tacit knowledge. It can, of course, always be questioned how valid this knowledge is. I'll let you, the reader, judge as you read.

Summary of the chapters

By chance, one can end up spending much of

a life in surroundings very different from one's own motherland. Under such circumstances, many pressing questions present themselves begging for answers. How do we communicate with people who speak a language we do not understand? How do we relate to a culture significantly different from our own? How do we understand the notion of time in a country undertaking rapid transformations at incredible – almost incomprehensible – speed? And, approaching a main theme of the book, how do we understand context when moving around in the field of architecture? Based on my 34 years of experience in activities related to China, this book intends to offer some reflections on such questions.

The book will in various ways focus on permanence and change, and on similarities and differences in our cultures, as it moves along two axes of content: 1) my main profes-

第八章提出了20世纪50年代的住宅区是不是文化遗产的重要组成部分，以及是否值得保留的问题。我还在此讨论了住房政策调控下的交通和住房区位间的关系。

第九章涉及我在2011—2013年期间教授中国学生的经验，如我所质疑，当忽视批判性反思时，是否会发生一种隐蔽的文化殖民？考虑到学习中文和学习西方语言的巨大差异，我讨论了心智能力是不是语言系统的产物。当一个外国教授遇到中国学生会发生什么？彼此会产生怎样的影响？谁学到的更多？

第十章，我用两个案例阐述了转型进程中利益相关者参与的困境。案例一是2008年唐大明宫遗址公园国际竞赛。案例二是2011年至2014年汉长安城宫殿遗址区的变化。这两个项目都位于西安，且在过程之初都有人居住在遗址区内。人们是如何被纳入转型过程的？他们对于变化做何反应？来自旅游业、文化遗产保护部门和房地产发展的压力，这些都是

影响着当地居民未来的巨大力量。在这里，关于发展策略选择及其对于社会文化的延续和改变的影响，我提出了自己的一些思考。

第十一章有关20世纪80年代以来快速城市化进程中的中国农村。我在此用很多照片讨论了几个现象：粮食生产、使用传统技术处理当地资源、窑洞、地方艺术，比较了乡村与城市的婚庆方式。通过一些案例，探讨了城市化对于乡村社会的影响，以及城乡相互作用的关系。

第十二章介绍了陕西柞水凤凰镇的发展，这是一个按照风水原则形成的历史古镇。这一章重点探讨的是小镇与自然和历史的联系。继而阐述了日益增长的旅游业的影响，以及由此带来的问题。一个小村庄的旅游业应如何把控？

第十三章重点介绍了另一个历史悠久的古镇——陕西陈炉制陶小镇及其从唐代至今制陶工艺生产发展

sional interest –inhabited, cultural heritage, and 2) my own personal and professional experience of China's development during this period. Two issues might have been included but are let out: During this period of 35 years, my Chinese colleagues and friends have developed their own projects and practices of very high quality. I found it too comprehensive to go into that. It would demand another book. At least. The second issue that I have avoided, is diving into the long and rich history and monumental architecture of China. I stick to daily life architecture and the people of our time.

Chapter 1 describes my first contact with China, a stay of seven weeks in 1985 studying vernacular architecture in its diverse geographical locations. My first curiosity about cultural differences was a result of this tour. An introduction to approaching China from a country far away.

Chapter 2 describes Xi'an as I experienced it in 1985 and my first contact with Chinese colleagues, a contact which later developed into a permanent relationship spanning 34 years. Some detailed observations of contact with colleagues and common people, as well as local conditions in and around Xi'an during our visit in 1985.

Chapter 3 handles my first long stay in Xi'an in 1989, and how I experienced life on the campus of Xi'an Jiaotong University (XJU). The chapter also describes the important start of the work in DTMD in the center of Xi'an old city, a work which later continued for some 20 years, when I revisited Xi'an four years later. Housing conditions both on the university campus where we lived, and in the poor housing areas in DMTD, formed a good basis for following the rapid development in China in the years to come.

Chapter 4 is an introduction to DTMD

的简况。我检视了人口结构的影响和对废旧材料的再利用问题。与凤凰古镇一样，目前当地在寻求通过旅游来带动发展。在一个陶瓷逐渐被玻璃和塑料制品取代的时代，制陶传统如何发展？

第十四章是关于资源友好型建筑学。这一章介绍了一些针对环境问题的研究项目，探寻了推动可持续发展的技术解决方案的潜力。我列举了三个项目：第一个是1993年西安城墙下的住宅项目，选址在城墙内，利用该地块的现有资源设计一个住宅区。第二个项目涉及遗产保护项目中的资源流动，是2002年在西安鼓楼回民区一个地块中修复与新建的对比。第三个项目是在2014年至2016年，东南大学和挪威科技大学在山西省平遥县侯冀村进行调研，从节能角度研究如何对具有重大文化价值的历史院落进行整修和保护。

第十五章对1985年以来几个方面的发展进行了总结，部分是基于整个时期的经验，部分是通过2018

年和2019年对中国的重新考察。主题是城市化、交通、污染、食品、贸易和文化。在与中国相关的各种各样的活动中，我尝试总结出这些年中所遇到的最为重大的变化，以期告诉读者，在充满活力快速发展的中国，一个外国人是如何亲身体验到永恒与变迁的。

在不同情形下人与人的会面意味着彼此目光的对视——正如本书标题"挪威建筑师镜头下的中国"所示，当然也会有从中国视角对挪威的观察。为说明这种关系，书中收录了董卫教授和王韬主编的两篇序言，他们都曾在挪威科技大学建筑学院攻读博士。还有一篇后记是对刘克成、肖莉教授的访谈，他们来自我在西安的主要合作机构西安建筑科技大学。从1985年到2019年，我和他们四人都曾长期合作。

除非专门说明，本书中所有照片和图纸均出自笔者本人。

in 1989, and to transformation processes going on there. The chapter also describes the first joint master's course between Chinese universities – this time in 1991 it was XJU, and the NTNU. The students worked on problems around inhabited historic housing in DTMD, investigating the potential and limitations of the existing, densely populated parcel system. The first studies of the housing conditions in DTMD in 1989, and later start of presenting the complex problems for European architecture students in cooperation with Chinese colleagues, gave an increased understanding of the situation there.

Chapter 5 delves into the role and challenges of professionals facing various types of cultural differences. Based on a Sino-Norwegian agreement, there was a contract for the development of a part of DTMD, comprehensive in terms of involved aspects. The project lasted for 6 years, from 1997 to 2003. The chapter

describes the situation of the NTNU team when it had to pass several cultural thresholds when working in Xi'an: Asian/Chinese/local Xi'an/Muslim. All within the framework of the nation-to-nation agreement.

Chapter 6 deals with how to relate to inhabited cultural heritage, also in DTMD between 1997 and 2003, and goes deeper into the reality of the district (a reality which is representative of similar cases elsewhere). I discuss doctrinal texts on conservation and maintenance and show how these texts relate to DTMD project. The increasing international interest in protecting vernacular housing milieus has evolved and been followed by Chinese cultural heritage authorities. Being loyal to the protection rules has been challenging for the residents of historic vernacular housing areas like DTMD, which is described in this chapter.

Chapter 7 deals with activities in public space, activities that seem to be stable over time, undergoing few changes. Playing mahjong, cards and chess, and park activities are illustrated by many pictures. In outdoor life the relationship between permanence and change is evident. Various activities have long traditions and are unchanged, while others are modernized and find new forms.

Chapter 8 asks whether the housing districts of the 1950s are an important part of the cultural heritage and if they are worth maintaining. I also discuss the relationship between traffic and the location of housing that is regulated by housing policy.

Chapter 9 refers to my experiences teaching Chinese students in the period 2011-2013, as I question if hidden, cultural colonization takes place when critical reflections are neglected. I discuss if mental abilities are a product of language systems, taking into consideration the vast difference between learning Chinese and learning Western languages. What happens when a foreign professor meets Chinese students? What kind of influence is that? Who is learning most?

Chapter 10 uses two cases to describe dilemmas with the involvement of stakeholders in transformation processes. Case one is the international architectural competition for the Tang Dynasty Daming palace in 2008. Case two is the transformation of the Han Chang'an City palace area of the Han Dynasty in 2011-2014. These two are both situated in Xi'an, and both were inhabited at the outset of the processes. How are people involved in the transformation processes? How do they react to the changes? The pressure from the tourist industry, the cultural heritage authorities, and the need for more urban development land – all are powerful forces affecting the future of the local residents. Here are some reflections on the choice of transformation strategies and their effect on sociocultural permanence and change.

Chapter 11 deals with the countryside of China in the rapid urbanization process that has taken place since the 1980s. I describe several phenomena – here also with many photos: food production, local resources handled with traditional techniques, cave dwellings (the housing type in the soil of the Loess Plateau in Shaanxi Province), local art, and finally comparing marriages in villages and in cities. The effect of urbanization in rural societies and the urban/rural mutual influence is discussed by some selected examples.

Chapter 12 describes the development of Fenghuang, Shaanxi Province – an old, historic town that has been planned according to *Feng Shui* principles. The focus is on the links to nature and to the history of the town. The influence of growing tourism and the problems this brings are then described. How to control tourism in a small village?

Chapter 13 has another old, historic town in focus, the pottery town of Chenlu, Shaanxi Province, and the development of pottery production from the Tang Dynasty till today. I look at the influence on the population and the reuse of waste materials. Presently, just like Fenghuang (chapter 12), they are looking for tourism to impact development. How are the pottery traditions developing when ceramic products are replaced by glass and plastic wares?

Chapter 14 handles resource-friendly architecture. The chapter describes some projects dealing with resource-friendly development, questioning the potential of boosting sustainable solutions. I list three projects: The first is from 1993 in Xi'an where the location of a housing area inside the City Wall was designed to utilize the resources of the site. The second deals with the resource flow in conservation projects where restoration is compared with new construction on sites in DTMD, Xi'an, in 2002. The third took place in 2014-2016, when the SEU in Nanjing and NTNU performed research in the Houji Village in Pingyao County, Shanxi Province, investigating how to refurbish and conserve historical courtyards of great cultural value-boosting in the perspective of energy efficiency.

Chapter 15 gives a summary of several aspects of development since 1985, partly based on experiences in the whole period, and partly by revisiting China in 2018 and 2019. The main themes are urbanization, transport, pollution, food, trade, and culture. In the aftermath of various activities related to China, I try to sum up the most substantial changes I have met during these years, hoping to inform the reader how a foreigner experience permanence and change in the rapid development and dynamics in China.

Meeting with people in various situations and circumstances means that – as the title of the book indicates (China through Norwegian architects' lenses), there will, of course, be glances the other way around. Norway through Chinese lenses, one could say. To illustrate this mutual relation, I have chosen to include two prefaces by Professor Dong Wei and Executive Editor-in-chief Wang Tao, both former Ph.D. candidates of our department at NTNU. There is also an afterword based on interviews with Professor Liu Kecheng and Professor Xiao Li, both

in our main partner university in Xi'an, XAUAT. With all four of them, I have had long-term collaboration during the period from 1985 to 2019.

All photographs and drawings in this book are my own work unless otherwise stated.

01

初访中国
1985

01

FIRST
VISIT TO
CHINA
1985

1985年，我很幸运地从所在大学获得了一笔旅行资助，用于研究中国不同地理环境中的风土建筑（即传统建筑）。我和妻子、儿子乘坐火车，从我们的家乡挪威特隆赫姆抵达北京，历时九天九夜。火车经西伯利亚铁路穿越苏联，那种远行的感觉非常强烈。从莫斯科向东，火车在一天一夜里跨越了15个经度。表现在时区的变化上，手表需要每天向前拨一小时，于是每天变成了23小时。这对早起的鸟儿来说也许是理想的，但对夜猫子来说恐怕是个麻烦，因为对他们来说晚上本来就总是太短。这样前往中国的方式是一个有益的提示——两国间的物理距离是如此之远，那么文化上的差异呢？会不会和地理距离一样巨大？这于我，成为今后数年一个亟待回答的迫切问题。＊1

火车旅行从苏联到中国的转换很有趣。在苏联一侧的边境上，有一次火车停了个把小时，车厢被抬到适合中国一侧铁道轨距的转向架上。而我们临时待在一个人满为患的地方，房间阴暗，几把椅子上铺着米色的布，工作人员（我记得他好像生着闷气）送来用灰色纸包着的雪糕。转换完毕，我们重新上车，火车开始行驶，几分钟后驶入中国边境一侧的车站。迎接我们的是一个精心布置的到站仪式，彩灯闪烁，喇叭里传来清亮的维也纳华尔兹舞曲，站台上排着微笑鞠躬的工作人员，穿着洁白的、刚熨好的制服。苏联和中国都是社会主义国家，但两者的差别此刻清晰地显现出来。

火车穿过黑龙江省继续向北京进发，透过车窗，我们可以看到公路和田间地头的人们在活动，做着我们很容易理解的活计。从这种浅表层面，我们似乎可以根据自己的文化背景解读所看到的一切，跨文化理解这件事情看起来似乎并不难，但当我们走下火车走近这片土地和人的时候，又会发生什么呢？只有时间能够告诉我们。

1985年，在中国，外国人原则上是可以独立出行的。当时负责接待游客的部门明显缺乏经验。虽然，有一个专门的组织——中国国

＊1　为了介绍我在中国参与的各种活动，本书头几章讲述了我是如何带着自己的期待和偏见慢慢去接近中国文化。我希望读者不要将其看作一种自我中心的表达，而是一种坦诚的努力，以期读者可以对本书进行批判性的阅读。

际旅行社，但会说英语的工作人员非常少，帮助游客的能力也很有限。游客当时须使用一种价值比人民币高15%的特殊旅游货币"外汇兑换券"。外国游客消费要以外汇兑换券付款；使用这种货币可以进入专门的"友谊商店"，在那里能买到当地商店没有的物品。

当时中国的旅游产业仅有数量有限的外国游客和寥寥几辆大巴车就能装下的国内游客，与今天中国的旅游业形成鲜明的对比——无限制的行动自由、无须特殊货币、功能完善的旅行服务、气派的银行大楼、遍布各个角落的自动取款机、会说英语的酒店工作人员、定期的航班服务，以及可以轻松追赶法国 TGV 和日本新干线的高速列车，更别忘了巨大的旅客流量。比起当年，现在有了众多的外国游客，但更引人注目的是，中国大陆本土游客的庞大数量，他们现在经济上有实力，而且政府也鼓励大家旅行。

在中国境内，我们主要乘火车旅行，这在当时意味着蒸汽机车和拥挤的车厢，特别是格外拥挤的硬座车厢。对于一位不懂中文的外国人来说，很难理解中国的售票系统。此外，我们还拿到了一张清单，上面列出了外国人不需要特别允许可以前往的地方。我们计划想去的地方恰好都在清单上，因此，它对我们的旅行没有产生任何影响。

由于时间充裕，加之联系人的帮助，我们的自助旅行还算顺利。我们的行程可以简单地总结如下：北京的胡同街区，呼和浩特与内蒙古草原的蒙古包，兰州的现代城市风土建筑，敦煌莫高窟及沙漠住居，西安的城市院落建筑，苏州园林与运河民居，绍兴的稻田和临水民居，最后是上海的里弄街区。

总的来说，对于我们这是一次很好的认识中国的旅行。在世界任何地方，目睹人们如何在各种苛刻条件下生活居住，都令人敬畏。正是这种魅力主导了我们的旅行；我们对人们在何处定居、如何定居、如何安排生活、如何充分利用当地潜力，以及如何获得有限的资源充满钦佩。

图 .01-1
穿越戈壁
（1985年）

Figure.01-1
Gobi desert
(1985)

In 1985, I was fortunate to receive a travel grant from my university, to study vernacular architecture – i.e., traditional buildings – in diverse geographical settings in China. My wife, my son, and I traveled mainly by train for 9 days and nights, from our hometown Trondheim in Norway to Beijing. Through the Soviet Union, we rode the Trans-Siberian Railway. The feeling that I was traveling far away was really strong. From Moscow and eastwards, the train passed 15 degrees of longitude in one day and one night, underlined by the systematic change of time zone, which went forward one hour every day. Each day had 23 hours, which is perhaps ideal for an early bird, but probably troublesome for a night owl, for whom the evenings are always too short. The journey turned out a useful reminder that there is a long physical distance between the two countries. What about the cultural distance? Would there be differences congruent with the geographical distance? That became a burning and exciting question to grapple with in the years to follow. * 1

During the train journey, the switch from the Soviet Union to China was interesting. On the Soviet side of the border, there was a stop for a couple of hours to lift the train wagons over to bogies adapted for the gauge on the Chinese side. We stayed in an overpopulated spot, a gloomy room with a few chairs covered with beige fabric. Blocks of ice cream wrapped in grey paper were served by the staff (a rather sulky staff, too, as I remember it). The bogie swap finished and we re-boarded the train, the train started

---

*1   In order to illustrate the background for the activities in China in which I have been involved, the first chapter describes how I gradually moved into the Chinese culture with those expectations and prejudices that I brought along. I hope the reader will read this not as an expression of ego-centrism, but rather as an attempt at transparency and encouraging critical reading.

moving, and a few minutes later rolled into the railway station on the Chinese side of the border. It was a well-staged arrival, illuminated by colored lanterns, crisp Vienna waltzes over the loudspeakers, and a smiling and bowing staff lined up on the platform, in clean, white, freshly ironed uniforms. The Soviet Union and China, are both communist states. The contrast between the Soviet and the Chinese side was clearly spelled out.

Continuing towards Beijing through the Heilongjiang Province, we could observe, through the train windows, people on the roads and in the fields busy doing what we easily understood with some relevance and importance. On this superficial level, it seemed possible to interpret, against our own cultural backdrop, what we saw. This was promising. But what would happen when we moved closer to land and people? Only the future would tell.

In principle, one could travel around independently in 1985. At that time, the departments and institutions handling tourists were conspicuously inexperienced. Certainly, there was an organization for the purpose, CITS (China International Travel Service), but English-speaking staff was rare and had limited capability to help tourists. Travelers had to deal with the special tourist currency, FEC (Foreign Exchange Currency), valued 15 percent higher than the national money, RMB (Renminbi, the people's money). Tourists were required to pay with FEC; this money also gave access to the so-called Friendship Stores, where customers could buy goods not available in the local shops.

This tourist realm, with a limited number of foreign tourists and a few busloads of Chinese travelers, is in sharp contrast to the activities and arrangements of today. Unlimited freedom to move, no special currency, a well-functioning travel service, numerous grand bank buildings, ATMs on every corner, English-speaking hotel staff, regular flight services, and high-speed trains that can easily keep up with the French TGV and Japanese Shinkansen. And let's not forget the enormous volume of tourists. There are more foreigners, of course, but more striking is the flow of mainland Chinese, now economically able - and requested by the government - to travel around.

Inside China, we mainly traveled by train, featuring steam engine locomotives and tightly packed wagons, particularly crowded in the Hard Seat class at that time. Ticket systems were difficult to comprehend for a non-Chinese-speaking foreigner. We were given a list of places we as foreigners could visit without asking for special permission. The list contained all the places we had already planned to see, so this regulation was no hindrance to our tour.

Thanks to abundant time, and thanks to helpful contact persons, we managed quite well to move around on our own. Briefly summarized, our travel route was: Beijing - with the *hutong* neighborhoods; Huhehot and grassland - with the yurts of the nomads; Lanzhou - with modern, urban vernacular; Dunhuang - with the Mogao grottoes and desert settlements; Xi'an - with its urban courtyard structures; Suzhou - with the gardens and the canal housing; Shaoxing - with rice paddies and settlements along water-ways; and finally, Shanghai - with the *linong* districts.

Altogether the tour offered a good introduction to vernacular China. It is awe-inspiring to witness, anywhere in the world, how people find ways to live under marginal conditions. This fascination dominated our travel; we were full of admiration for where and how people had settled, how they had arranged their lives, and how they made the most of local potential and the access to resources that were often limited.

02

西安
1985

02

Xi'an
1985

火车从塔克拉玛干沙漠中的柳园站出发到达西安需要36小时，车票没有固定座位，硬座车厢里非常拥挤。人们睡在行李架上、过道上、两节车厢之间的连接处。在这里，我第一次观察到一个现象，一个我在以后岁月里会不断遇到的现象：车门一打开，空荡荡的空间瞬即被填满。后来，我在中国的街头交通中也观察到同样的情况：自行车流如水流般移动，而行色匆匆的骑车人则在这股洪流中进进出出；或当汽车不管不顾地从垂直于主干道的支路上汇入，长时间堵塞交通；又或在另一种尺度下，当拥挤庭院中的住户蚕食公共空间的时候。对于一个来自通常情况下空间充裕、很少需要和别人分享空间的国家的人，我渐渐明白了生活在一个人口稠密的国家是什么情形。在去西安的火车上，我观察到一位没有座位的军人。他上车后第一个动作是把帽子放在两排座位中间的桌上。等了几个小时，桌子旁边一位乘客离开去上厕所，这名军人旋即在空位上坐了下来，等那人回来后，他又礼貌地让出座位。其间，这位军人与同桌乘客聊得火热，与他们更加熟络。又过了几小时，一位乘客下了火车，这名军人便理所当然地坐在了空出的座位上。

这趟旅程中，我们和一位在西藏做完田野工作回家的地质专业学生成为朋友。到达西安后，我们下了火车，他原本计划继续前进，但还是决定下车，仅为在西安陪我们一段时间。在火车站，一位人力车夫说带我们去车站附近的旅馆。他在不远处一栋非常破旧的房子外停下来，显然那是他的家，他进去让妻子从床上起来，以给我们腾出房间。我们拒绝了，他继而表示带我们去一家专业酒店。街上非常寂静，没有车水马龙的噪声，一些人在深夜中轻声私语。人行道上，一些人躺在从闷热拥挤的房间中搬出的床上睡觉。到了苏式纪念建筑风格的人民饭店，酒店前台已经上班，我们拿到房间，并提出给火车上结识的新朋友付房费。但他作为中国公民，不允许在接待外宾的人民饭店过夜，所以只好离开。我们约定次日见面，但到了第二天他没有露面。我不知道出于什么原因，也许他觉得被冒犯、尴尬，或是愤怒？我们很遗憾以这种令人不快的方式失去了一个可能的好朋友。

与中国同事和
西安的
首次邂逅

我们在北京结识的一位中国同事给了我们一份他大学同学的名单。正是通过这个名单，我们在西安联系上一位在陕西省建筑设计院工作的建筑师，并在那里受到了热情招待。对于一位挪威人来说，这个设计院360名员工的规模让人印象深刻，因为挪威的建筑师事务所普遍很小，通常只有5~10名员工。我们是第一批参观这个设计院的外国人，他们为我们安排了一辆汽车，让我们可以四处看看城市周边的乡村。他们还宴请我们，席上有无数美味佳肴。此外，他们还举办了一个研讨会，请我们即兴演讲。我妻子介绍了挪威，我则被要求谈谈摩天楼的问题，而我既无任何这方面的工作经验，也无相关知识。我转而谈了谈其他流行的现代建筑样式。我们俩在黑板上用粉笔绘图解释，并得到一位翻译协助，这位聪明的翻译后来成了我们的好友。四年后，这位翻译帮助我们获得了中国大学的邀请，让我们在中国度过一段学术休假时光。

在一次乡村旅行中，我们看到有一些农民并排在路边卖西瓜，就停下向其中一位农民买了西瓜，结果引起了旁边一位农民的强烈不满，很严厉地问我们为什么不买他的瓜。我们的司机不经意间说了一句"因为他家瓜更甜"，结果引来一场激烈的口角，我们只好用钱安抚被冒犯的农民，以便能继续上路。此前我们遇到的人都是彬彬有礼、乐于助人的。这一幕生动揭示了普通人的另一面，有时候辱慢他人是多么容易，人受辱后又是多么冲动。

当时西安城中心主要是四至六层高的楼房、低层建筑、四合院和少量面向街道的商铺。大街上鲜有外国游客，因此我们出现在哪儿都会引起强烈关注。城里没有路灯和广告霓虹灯，夜晚一片漆黑。没有商店橱窗照亮街道，晚上的街道很寂静，几乎空无一人。清晨，人们通常穿着睡衣，提着要倒的尿壶和垃圾桶去公共厕所。许多人在人行道边进行晨间盥洗——梳头刷牙。沙发睡具被抬回屋内。然后，小商贩们开始登场，用尖厉的叫卖声和铃声兜售商品。此后，随着白天温度逐渐升高，人们在酷暑中用湿毛巾围着脖子降温，坐在低矮的木凳上，喝着人行道上小贩售卖的用冰块冰镇的瓶装汽水或茶水。

人民饭店附近可以租到自行车。城市的交通状况尚可，路面上主要是公交车，极少的出租车、货车、骡马车，还有无数的自行车。我们从城墙的南门出城，往小雁塔方向去，然后再前往大雁塔。一路上我们经过村庄和农田。今天，看看这一带的繁华喧闹的城市环境，和1985年我们初到这里相比，发生的变化让人瞠目结舌（见第15章）。

图 .02-1
西安北院门
（西安，1985年）

Figure.02-1
Beiyuanmen
Street
Xi'an (1985)

The train needed 36 hours from departure from Liuyuan in the Taklamakan desert to Xi'an. There were no seat reservations, the Hard Seat compartment was crowded. People were sleeping on the luggage racks, in the corridors, and on the links between two wagons. Here I first observed a phenomenon that I would encounter over and over in the years to come: The filling of empty space as soon as one opens up. Later I would, for example, observe the same in street traffic, where the stream of bicycles moved like a flow of water, and bikers who were in a hurry moved in and out of this stream. Or when cars moved perpendicular to the main street, blocking traffic for long times. Or - on another scale - when families in congested courtyards nibbled public space. Coming from a country where generally spoken there is abundant space, and where few people need to share space, this has illustrated what it means to live in a densely populated country. On the train to Xian, we observed a soldier who did not have a seat. His first move was to put his cap on the table between two rows of seats. After waiting for a couple of hours, one of the passengers beside the table left for a toilet visit. The soldier immediately sat down in the vacant seat and left it politely when the original seat holder returned. During that time the soldier chatted lively with the co-passengers, getting better acquainted with them. Another few hours later, one of the seated passengers left the train, and the soldier then was the self-evident person to take over the free seat.

During this journey, we became friends with a geology student going home after a fieldwork period in Tibet. When arriving in Xi'an we left the train, and even if he originally had plans to continue further, he decided to do the same, just to accompany us for some time in Xi'an. At the railway station, a man on a rickshaw offered to take us to a hotel close to the railway station. He stopped outside a very shabby house nearby, evidently his home, and went in and told his wife to leave her bed in order to make room for us. We declined, so instead, he offered to take us to a professional hotel. The streets were very quiet. There was no traffic noise, just some people softly talking together in the dark evening. On the sidewalk, there were people sleeping in beds moved out from what were probably warm and crowded rooms inside the buildings. When arriving at the monumental, Soviet-style Ren Min Hotel, the reception was open, and we got our room and we also offered to pay for a room for our new friend from the train. However, as a Chinese citizen, he was not allowed to stay overnight at the Ren Min Hotel, and he had to leave. We agreed to meet the next day, but the next day he did not show up. It is hard to know how he felt. Offended, maybe. Embarrassed, angry. We felt sorry to lose, in such a disagreeable way, what could possibly have become a good friend.

FIRST CONTACT WITH COLLEAGUES
AND THE CITY OF XI'AN

A Chinese colleague we met in Beijing had given us a list of former classmates from his university, people he told us to contact when we arrived at the respective places where they worked. In Xi'an, we got in touch with an architect who worked at Shaanxi Provincial Architecture Institute, where we enjoyed the benefits of their hospitality. The size of the office, with 360 staff members, was impressive for a Norwegian, who was used to small-scale architect offices at home, more commonly employing a staff of 5-10 persons. We were the first foreigners visiting this unit. They arranged a car so we could get out and see the countryside around the city. They also threw a big banquet with I do not know how ma-

46

ny tasty dishes, and a seminar where we were asked to give improvised lectures. My wife introduced Norway, and I was asked to talk about skyscrapers, of which I did not have any experience or knowledge. I dodged by talking about other popular and modern building types. We both made chalk illustrations on a blackboard and had assistance from an interpreter, a clever translator we became good friends with. Four years later this same translator arranged the invitation from the university where we stayed for a sabbatical period.

During the tour of the countryside, we stopped to buy watermelons from a farmer on the roadside. There were several people side by side selling their own melons, and a neighboring farmer asked harshly why we did not buy his melons. A thoughtless comment from our driver ("because his melons are sweeter") caused a loud and rowdy verbal fight, which had to be settled by appeasing the offended farmer with money so that we would be able to continue on our way. Until then we had only met polite and helpful people. This episode informatively revealed another side of common people. How easy it could sometimes be to insult someone, and how impetuously an insult might be followed up.

The city center was dominated by buildings 4-6 stories high and low-rise structures, courtyards with very few shops addressing the street. Foreign tourists were rare, causing intense attention everywhere we appeared. In the absence of streetlights and advertisements, evenings were dark. No shop windows threw light onto the streets, which were quiet and almost empty in the evening. In the early morning, people headed for the public toilets, often in their night gowns, carrying chamber pots and garbage boxes to be emptied. Many people did their morning toilet on the sidewalk–combing their hair and brushing their teeth. Sofas were removed and carried back to the house. Peddlers arrived announcing their commodities with shrill voices and bells. Later the air turned hot. People wrapped wet towels around their necks to cool down in the heat, and sat on very low wooden stools refreshed themselves with sweet bottle water stored on ice blocks or tea sold by vendors on the pavement.

Bikes could be rented near Renmin Hotel. Traffic was moderate: there were mainly buses, very few taxis, lorries, mules, and horses. And there were myriads of bicycles. Leaving the city through the city wall at the South Gate, heading for The Small Wild Goose Pagoda and further out the Great Wild Goose Pagoda, we passed villages and farmland on our way. Looking at the urban environment in this area today, the transformation that has taken place since we were there in 1985 is mind-boggling. (See chapter 15).

1985 年的西安印象

图 .02-2
可口可乐的到来
（西安，1985年）

Figure.02-2
The arrival of
Coca-Cola
Xi'an (1985)

图 .02-3
防尘茶杯
（西安，1985年）

Figure.02-3
Dust-
protected
tea-glasses
Xi'an (1985)

图 .02-4
公园图书馆
（西安，1985年）

Figure.02-4
A public park
library
Xi'an (1985)

图.02-5
路边餐食
（西安，1985年）

Figure.02-5
Side-walk
restaurant
Xi'an (1985)

图 .02-6
流动小贩
（西安，1985年）

Figure.02-6
Mobile
peddler
Xi'an (1985)

图 .02-7
货物运输
（西安，1985年）

Figure.02-7
Goods
transports
Xi'an (1985)

图 .02-8
农民从餐馆
收集泔水
（西安，1985年）

Figure.02-8
Farmer
collecting
food remains
from restaurants
Xi'an (1985)

油毡·油友
云南厂家供应
铁红黄黑绿兰
咸宁东路船务

图.02-9
清扫路面
（西安，1985年）

Figure.02-9
Cleaning
the street
Xi'an (1985)

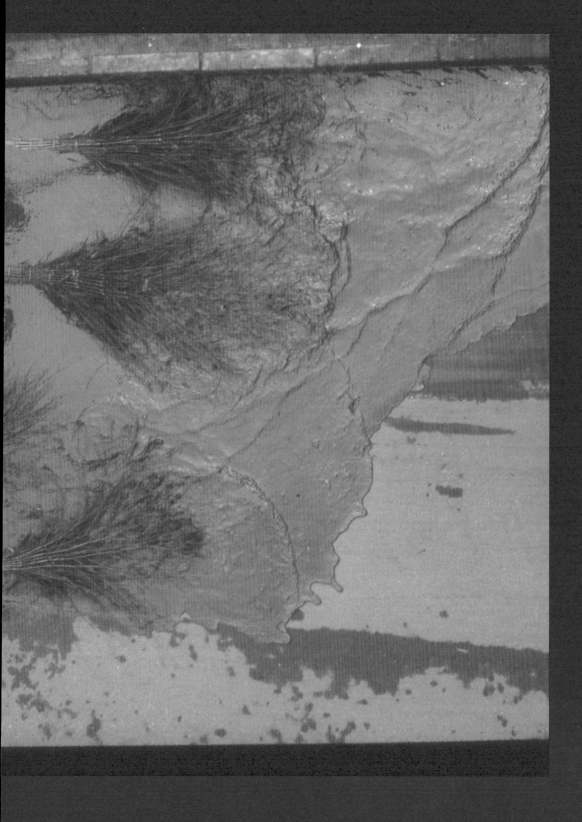

图.02-10
菜市场
（西安，1985年）

Figure.02-10
The vegetable
market
Xi'an (1985)

图.02-11
自行车修车铺
（西安，1985年）

Figure.02-11
Bicycle repair
shop
Xi'an (1985)

图.02-12
粪水收集
（西安，1985年）

Figure.02-12
Manure
collecting
Xi'an (1985)

03

西安
1989

03

Xi'an
1989

1989年，为期一年的学术休假让我再次来到中国，这次时间更长，而且是和我妻子、我们一岁半的女儿一起。原本的计划是去1985年访问过的陕西省建筑设计院，在那里做建筑师的工作。但新任院长不支持让外国人到设计院工作，我们只好请翻译老友帮忙另谋他途。最终，我们在西安交通大学待了几个月。该校建筑学院一位年轻有活力的院长发出邀请，在这里，我以访问教授身份工作，部分工作是讲课做研究，部分是合作参与国际建筑竞赛，为成立不久的建筑学院的发展贡献力量。

从北京来的飞机降落在西安老机场*2，唐代时，这里是长安西市所在地，那时的西安是丝绸之路的起点，可见其繁华。现今，往日辉煌已经无迹可寻，这里只是一个小小的简陋的省城机场，乘客下飞机后要在一个小铁皮棚子里提取行李。数年后，西安机场搬到了城市西北地势开阔的地方。起初，新建机场打着"文明机场"这个奇特的名称，航站楼最初的规模和航班数量都非常有限。但是如今，西安机场拥有三个航站楼，既服务于国际航线，也服务于覆盖广泛而繁忙的国内航线。机场的变化也是过去25年这个国家在许多领域实现现代化的一个生动例子。

我们抵达西安时天气有些沉闷，乌云低垂，下着雨，气温只有零上几度。等候我们的是面带微笑的大学校长和年轻的院长，两人穿着中山装欢迎我们的到来，并安排了前往学校招待所的车辆，整个逗留期间那里将是我们的居所。

校园生活

西安交通大学的"饮水思源"铭言，让我们觉得非常有感染力，我们也很快适应了当地的生活条件。我们有一个大房间（我们2月抵达时屋子热得像蒸桑拿，3月集中供暖结束时则像冰窖），我们有自己的卫生间，在宾馆餐厅吃饭（所有的饭菜都有生姜的味道——显

---

*2　译者注：西安老机场，即西关机场，建于1924年，位于西安市安定门外西郊。作为西安市的老机场，西关机场曾是中国离城区最近的机场，距西安市中心仅四公里。

然厨师很喜欢这种风味）。我们白天工作的时候，女儿送去交大幼儿园照看（她在那儿被老师宠坏了，每当她发出一些或被误认为是中文的声音时，老师就奖给她一块巧克力，以致下午很难说服她吃晚饭）。宾馆的附近有邮局（邮票必须用柜台上一个生锈的盒子里的胶水粘在信封上）、电信局（打电话和使用传真机），房间里有电视。所有一切都让我们觉得舒适，处处很受照顾。

工作起始，我们被礼貌地邀请作为团队成员参与一个美国密尔沃基的国际规划竞赛。我们希望在有限的时间内尽可能多地了解中国，所以我们表示做一些与中国当地情况直接相关的事似乎更有意义，这当然被接受了。因此，除了偶尔参与讨论，我们并未在这个竞赛上投入太多时间。不过，这也让我们看到了中国同事们对现代建筑和城市发展的思考。

回归基本生活——
西安鼓楼回民区

在西安首次逗留的五个月期间，我第一次接触到位于西安市中心的鼓楼回民区。这是我与该片区长久关系的开始，此后多年中，这里成为各种相关研究合作活动的现场。自其初始，清真寺周边那些狭窄街道上上演的各种生动的当地生活内容就非常令人兴奋。与我妻子和中国同事们一道，我开始了对该片区的整体调查，不得不说，这里情况的复杂性让人着迷。年轻的建筑学院院长向我们提出，这片邻近西安市中心明代鼓楼的区域是一个可能的研究对象。这是一片气质独特迷人的区域，居住着约三万穆斯林，有着生动鲜活的生活氛围，有着热气腾腾的街边餐馆，这其间还散布有十座清真寺。白天人山人海，晚上则一片寂静。起初，我们的关注点集中在低劣的居住条件上，但在随后数年，文化遗产渐渐成为更重要的课题。

研究项目的具体目标是帮助提升当地居民的居住质量，并保护西安最重要的清真寺——化觉巷大清真寺周边的城市肌理环境。同时，这个项目的另外一个目标是发展西安交通大学与我所在大学之间的联系，当时叫作挪威理工学院（NTH），即今天的挪威科技大学（NTNU）。

74

回坊的氛围在几个方面吸引了我，低劣住房条件下的居民生活所表现出的体面尤其令人印象深刻。对于在非常不同条件下长大的我来说，这在某种程度上是生活的返璞归真，也是工作的挑战。什么是体面生活的最低居住标准？此后几年中，当我们与学生、研究伙伴们一起在那里工作时，这个问题反复出现。

图 .03-1
校园书摊
（西安，1989年）

Figure.03-1
A bookstall
on campus
Xi'an (1989)

图 .03-2
西安交通大学
"饮水思源"碑
（西安，1989年）

Figure.03-2
When drinking
water, think of
the source
Xi'an (1989)

图 .03-3
学生宿舍
（西安，1989年）

Figure.03-3
Student
dormitory
Xi'an (1989)

图 .03-4
大清真寺
（西安，1989年）

Figure.03-4
The Great
Mosque
Xi'an (1989)

图 .03-5
拥挤的院落
（西安，1985年）

Figure.03-5
Congested
courtyards
Xi'an (1989)

图 .03-6
严格控制的用水
（西安，1989年）

Figure.03-6
Water was
scarce and
strictly
controlled
Xi'an (1989)

A year of sabbatical leave from my university allowed me to visit China again for a longer period and together with my wife and our one-and-a-half-year-old daughter. The original plan was to stay at the Shaanxi Provincial Architect Institute, which we visited in 1985, to work there as architects. The new director, however, did not support the idea of having foreigners working there, so with the help of our old friend the translator we had to find other options. We ended up staying for a few months at XJU. Here I worked as a visiting professor. A young and dynamic dean of the school of architecture at the university invited me, partly to give lectures, partly to co-operate in an international architect competition, initiate research, and contribute to the development of the recently established school of architecture.

The plane from Beijing landed at Xi'an airport in the area where, during the Tang Dynasty, the West Market was located. Back then Xi'an was the terminal of the Silk Road. Now there were no traces of the glorious past, just a small, provincial airport where the luggage could be found in a tiny tin shack. Years later, the airport was moved to the Xianyang area, northwest of the city. Brandishing the curious name "Civilized Airport," the terminal was initially limited in size and volume of flights. Today the airport has three terminals serving international flights as well as a comprehensive domestic network. This is a good example of the modernization that has taken place in the country, in so many areas, during the last 25 years.

The weather at our arrival was a bit dreary. Low clouds, rain, and a temperature just a few degrees centigrade above zero. Waiting for us were a smiling president of the university and the young dean, both dressed in Mao suits. They wished us welcome and accommodated transport to the university guesthouse, where we would live during our entire stay.

CAMPUS
LIFE

Charmed by the slogan of the university, "When drinking water, think of the source," we soon adapted to the local conditions of living. One big room, (hot like a sauna when we arrived in February, ice cold when the central heating was turned off in March), with our own bathroom, meals in the guesthouse restaurant, (all meals with a distinct taste of ginger – the kitchen evidently preferred that flavor), our daughter in the kindergarten (where she was spoiled by the teachers who gave her a piece of chocolate each time she uttered some sound that could be mistaken for Chinese – it was hard to persuade her to eat dinner in the afternoons), post office (where stamps had to be glued on the envelope by glue from a rusty box on the counter), telecommunication office (for phone-calls and using a telefax machine), TV in the room. Altogether we felt comfortable and well taken care of.

We were asked politely to be part of a team working on an international planning competition in Milwaukee, USA. Wishing to learn as much as possible about China while we stayed there, we expressed that it seemed to us more meaningful to do something directly related to local situations, and this was accepted, of course. Consequently, we did not invest much time in the competition, apart from occasional discussions. This gave, however, a view into how our colleagues reflected on urban development and modern architecture.

BACK TO BASICS –
HOUSING IN THE DRUM TOWER MUSLIM DISTRICT

During my first five-month stay in Xi'an in 1989, I was first introduced to the DTMD in the city center of Xi'an. It was the start of a long-term relationship with this area, the scene of manifold activities over many years. From the very beginning, it was very exciting to be confronted with the vivid, local life in the narrow streets surrounding the mosques. Together with my wife and Chinese colleagues, we started overall surveys in the area, much charmed, I have to admit, by the complexity of the situation. The young dean introduced us to the district near the Ming Dynasty Drum Tower in the city center as a possible research object. At that time it was a very exotic area inhabited by around 30,000 Hui nationality Muslims, with a vivid milieu, steamy street restaurants, and 10 mosques. It was crowded in the daytime, and silent in the evenings. At first, our attention was focused on the poor housing conditions, but in the years to follow, cultural heritage played a more important role.

The specific goals of the project were to contribute to improving dwelling qualities for the tenants and to safeguard the environmental urban fabric surrounding the Great Mosque, the gravity point, and the most important mosque of Xi'an. With this project, the aim was also to develop and strengthen the links between XJU and my university in Trondheim, The Norwegian Institute of Technology, NTH. (Later the NTNU).

The atmosphere in the Muslim District appealed to me in several ways. Not least impressive was the dignity displayed by the residents who were living in poor housing conditions. It was somehow back to basics and challenging for me who grew up under very different conditions. What is a minimum housing standard for a decent life? This question repeatedly turned up, in the years to come, as we worked there with our students and research partners.

04

面对
复杂性
西安
1989—
1996

04

FACING
COMPLEXITY
Xi'an
1989-
1996

起始阶段

为了充分了解鼓楼回民区的一般住房标准，我们制作了访谈表，分发给各家。结果很失败，因为没人填写和返还访谈表。在和一些居民交谈中，我们发现很多男性家庭成员都是文盲，无法填写表格。在学生们的帮助下，这些家庭最终回答了我们的问题，使我们成功收集到了需要的数据。

在走访该街区时我们采用的调查方法是社会人类学的个人观察法和非正式访谈。我们与居民、街道居委会、大清真寺的阿訇聊天，研究西安市规划局提供的材料，通过向所有家庭发放调查问卷收集信息。问卷表格的内容非常详细，包括家庭条件、住房环境的总体状况、对目前住房条件的评价、对未来房屋改善的期望等问题。调查结果以中、英文报告的形式总结。我们的基本印象是，这里居住水准非常不尽如人意。这份报告对我们后续几年的活动都有帮助，鼓楼回民区是中国和挪威学生开展各种城市更新课程的所在地，对我们后来的鼓楼回民区项目也颇有助益（见后续章节介绍）。

如上所述，我们首次调查的重点是鼓楼回民区的住房条件，在1990年，这里的条件可以说非常不尽如人意。人均2.7平方米的居住面积，只有少数院落有自来水，公共水龙头很少，做饭在室外，有为数不多的公厕。由于旧城改造涉及复杂的矛盾和关系，该片区住房条件的改善整体落后于西安的其他区域。

当时人们又如何看待这样糟糕的住房条件？一位老太太说，在这样的情况下不可能过上体面的生活。有些人想方设法把自己的私人领域扩展到院子中本已狭小、尚未被使用的公共空间。一位中国同事称之为"蚕食公共空间"。这种做法会在院子里引起纠纷，所以只能缓缓图之。其漫长过程如下：首先在屋外存放一些物品，比如炊具、砖头、几根木梁、自行车轮子，等等。接着，可能一年后，在存放的物品和院子通道间建一堵墙，最后在顶部加上屋顶。瞧！家里就有了一个新房间！这样的扩张逐渐使得庭院变成狭窄的通道，很多人告诉我们，由此常引起住户间无休止的争吵。

为了更好说明住房条件，我想先说一个我们在西安头几年所提的建议。这个建议是关于如何为住户提供更好的卫生条件。在勘察院落时，我们发现，在两个地块交界的地方，往往会有一小片未被使用的地块。如果能把这些小空间利用起来，连接在一起，就可以形成一个连续的区域，我们称之为"内部服务通道"。在这里可以修建基础设施，也可以设置供这一区域内所有地块使用的公共浴室和洗衣房。这个想法也是为了使家庭内部的活动不被游客干扰注目。这会是一个最低限度提高卫生标准和改善基础设施的解决方案，而且费用不高。不过，这个建议很快变得无关紧要了，因为仅仅几年后，庭院的整体条件就得到了改善，这也恰恰说明了这片城市区域的快速变化。

硕士生联合课程

自1991年起，我所在的挪威科技大学和中国一些大学的建筑系开始长期合作。第一次合作尝试是和西安交通大学，后来与清华大学、东南大学、西安建筑科技大学都建立了良好的合作关系。秋季学期一开始，老师和同学们在选定的地点进行田野考察。学期其余时间，学生们在各自的大学里利用收集到的数据进行规划和设计。

我还记得课程开始时内心的融融暖意。我们坐在教室中，中国学生坐在一个角落，挪威学生坐在另一角落，大家都有些害羞，同时又充满了期待。老师们也有些紧张，这是我们初次如此尝试。一番开场白后，所有学生被邀请混在一处。两个背景迥异但兴趣相近（建筑）的人群的会面非常美好。所有人立即开始用挪威语—英语和中文—英语聊起来。尽管他们都是20~25岁的成年学生，却像幼儿园里的孩子一样随性激动。

第一门课和此后所有课程都涉及城市和乡村老旧住区的发展。研究人们日常居住和生活的旧住宅区，研究如何使现代生活适应古老的建筑，反之亦然——使老建筑适应现代生活。这意味着一系列有趣和重要的问题。

国际规范和理论文本强调两种文化遗产：有形的，即可以直接触碰的；无形的，即不能直接触及的。更笼统来说，有形的（遗产）可以理解为我们的感官所能体验到的东西，而无形的（遗产）则是与一个社会中人们心中的传统相联系的东西。这也可以表述为文化与文化表现形式的区别。文化可以理解为人们在特定社会范围里思想意识中共同的东西。文化表现形式则是这种共识可感知的产物，例如，建筑、音乐、舞蹈、美术、园林、手工技艺产品、口头文学，等等。这些概念常常被混为一谈，但于我而言，这些区别对理解这些现象是有帮助的。有形的文化表现形式和无形的文化是相互交织的现象，在有人生活的历史地段形成文化遗产的整体，突显了生活的连续性，现在是过去与未来之间的一个阶段。有机会去关注这些历史街区的发展，为我们的思考和反思提供了丰富的素材。

中国的历史住区往往有高密度和居住标准不高的特点。对于生活在低密度住区且通常住房标准较高的挪威人来说，这些情况是一个挑战，一个非常幸运且有益的挑战。这为我们提出了几个问题：人们如何在这样的居住环境下保持生活的尊严并适应这样的环境？人们如何协商使用现有资源？我们尤其特别感兴趣的地方是，对该片区居民来说，什么是他们认为的文化遗产？他们是否认为文化遗产是重要的？以及他们是否尊重这种文化遗产？

通过对现有住人院落的调查，我们对历史建筑技术、老建筑的建筑形式、社会状况、文化的延续性（或不延续性）、住房标准都有了一定的认识。第一年的设计任务是如何改造一个单独的院子，或研究将两个或多个院子合并改造的可能性。

早期风土建筑研究认为，气候条件在其中起着至关重要的作用，受此启发，最初课程的重点是研究苛刻住房条件下能够和气候相适应的建筑。很快，随着关注兴趣范围的扩大，我们越来越意识到需要了解设计的文化语境。我们在鼓楼回民区的第一个任务即充分显示出这种情况的复杂性，气候只是寻求好的设计解决方案的几个因素之一。我们很快清楚地认识到，如果我们把自己局限在气候适应性的问题上，我们就会错过一个绝佳的机会，即通过一种整体的方法来面对设计的种种挑战，以从中学到更多的东西。

最后，我们制订了联合教学计划，把三门课程结合在一起：建筑学与人类学（基于半结构化访谈）、测绘记录与分析（包括测量历史建筑），以及建筑设计（基于另两门课程的基础材料）。自1995年起，由于采用这一课程模式，我们与新伙伴西安建筑科技大学建立了牢固的合作关系。

除了少数例外，我们和西安建筑科技大学开展的硕士生联合课程，主要是关于城市居住区。中国的传统住区往往位于城中心。因此，随着城市的快速发展，这些区域已经或有可能被拆除。城市中心区对房地产开发公司是富有吸引力的，很多低层历史住区通常会为现代高密度城市住宅让路。对学生们而言，面对这些区域的保护修复、提升和现代化之间的矛盾非常有教育意义。另外，我们的课程选择的一些传统历史院落，分别位于鼓楼回民区、碑林周边地区和鼓楼以南的正学街。前两个片区在城市总体规划中的"保护区"范围内，后一个则不在。"保护区"这一概念也充满疑惑。什么应该受到保护？应如何保护？为什么要保护？这些对任何人来说都是有挑战性的问题，是一个进行批判性思考和反思的良好起点，其成果应该反映于设计。作为一个在住房分散的居住环境长大的挪威人，生活在这里的人们的体面让我印象深刻。人们早上出门上班，从相当破旧的物质居住环境中走出来。衣着整齐，干净利落，面带微笑。

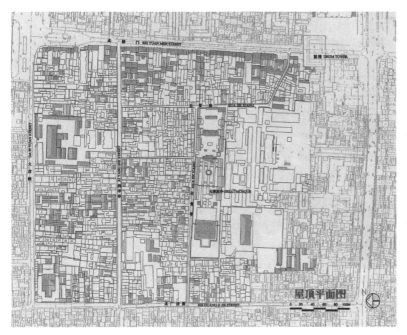

图 .04-1
西安鼓楼回民区
总平面图
（图源：西安鼓楼回民区挪威
团队工作报告，1994年）

Figure.04-1
Master Plan
of Xi'an Drum
Tower Muslim
District
(Photo credit: The DTMD
Report, NTNU, 1994)

图 .04-2
图中男子
拿反了问卷
（西安，1989年）

Figure.04-2
The man
held the
questionnaire
up-side down
Xi'an (1989)

图 .04-3
口头访谈
（西安，1989年）

Figure.04-3
Oral interview
Xi'an (1989)

图.04-4
街道生活
（西安，1989年）

Figure.04-4
Street life
Xi'an (1989)

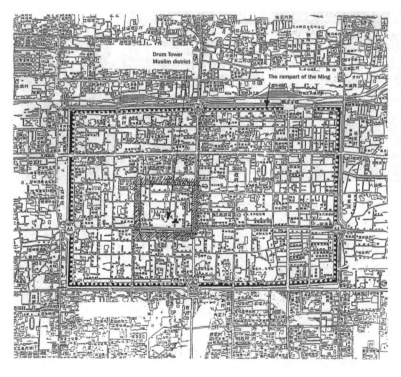

图.04-5
西安鼓楼回民区
区位图

（图源:《西安：现代世界中的
古老城市——城市形态的演
化1949~2000》Éditions
Recherches, 2007.）

Figure.04-5
The location
of the Drum
Tower Muslim
District

(Photo credit: *Xi'an-an Ancient
City in a Modern World*, 2007)

图 .04-6
拥挤的院落
（西安，1989年）

Figure.04-6
Congested
courtyards
Xi'an (1989)

图 .04-7/8
蚕食院落空间
（西安，1989年）

Figure.04-7/8
Nibbling
public space
Xi'an (1989)

图.04-8

Figure.04-8

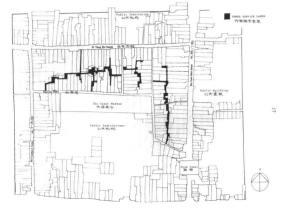

图 .04-9
内部服务通道

（图源：西安鼓楼回民区挪威
团队工作报告，1994年）

Figure.04-9
The inner
service lane
project

(Photo credit: The DTMD
Report, NTNU, 1994)

图.04-10
现场测绘
（西安，1998年）

Figure.04-10
Survey work,
Xi'an (1998)

（右页）图.04-12
鼓楼回民区四个
院落的合并改造
方案——1990
年挪威科技大学
学生作业

(Right)
Figure.04-12
Student project
in 1990 –
combining four
courtyards for
restoration
(Photo credit: The DTMD
Report, NTNU, 1994)

The Great Mosque

Drum Tower

图.04-11
大清真寺周边的
高密度住区图，
黑色部分为建筑
实体，白色部分
为露天空间
（图源：《西安：现代世界中的
古老城市——城市形态的演
化1949~2000》Éditions
Recherches, 2007.）

Figure.04-11
High density–
map of the
district
surrounding the
Great Mosque
(Photo credit: Xi'an–an Ancient
City in a Modern World, 2007)

Existing situation 1:200
现状平面 1:200

Plan ground floor 1:200
底层平面 1:200 （方案）

Plan first floor 1:200
二层平面 1:200 （方案）

图.04-13
鼓楼回民区
院落鸟瞰
（西安，1998年）

Figure.04-13
Aerial view of
the courtyards
in DTMD
Xi'an (1998)

图 .04-14
鼓楼回民区街景
（西安，2008年）

Figure.04-14
Street view of
DTMD
Xi'an (2008)

图 .04-15
碑林周边地区
（西安，2009年）

Figure.04-15
The Beilin
District
Xi'an (2009)

图.04-16
正学街
（西安，1998年）

Figure.04-16
The sign-
painter street
Xi'an (1998)

Initiating phase

To identify the general housing standard of the DTMD, we created interview forms and handed them out to the families. This turned out a failure because no one answered. Talking with some residents, we discovered that many of the male family members were illiterate and unable to fill in the forms. With help from students who assisted the families in answering our questions, we eventually managed to collect the data we needed.

The methods we used for the investigations were personal observations when visiting the district and informal conversations with residents, conversations with Street Committees, conversations with the Imam of The Great Mosque, studies of material procured by Xi'an City Planning Bureau, and collection of information using questionnaires distributed to all families. The latter was very detailed, including questions on family conditions, the general situation of the housing environment, evaluation of the present housing conditions, and hopes for improvements in the future. The investigations were summarized in reports in English and Chinese. Our main impression was that the housing standard was considered very unsatisfactory. This report became useful for our activities in the years to come, when the DTMD was the site of various project tasks for Chinese and Norwegian students, and finally for our DTMD project (which is described in later chapters).

As mentioned, the focus of our first investigations was the housing conditions, which were very poor. The floor standard was 2.7 square-meter per person, and only a few courtyards had access to water – public water taps were few, cooking was an outdoor activity, and a few public toilets were accessible. For different reasons, the area had lagged behind the general up-grading of housing standards in Xi'an.

How did people relate to the poor housing standard? An elderly lady said it was impossible to live a decent life under these conditions. Others tried to find ways of extending their private sphere onto the already small, unused public space of the courtyard. A Chinese colleague called it "nibble public space." This practice caused discussions and quarrels among the families of the courtyard, so one had to make it a gradual process. A normal procedure was as follows: First store some goods outside the wall of the dwelling, things like cooking utensils, bricks, a couple of wooden beams, bicycle wheels, etc. Next – maybe a year later – build a wall between the stored objects and the courtyard passage. And finally lay a roof at the top. Voilà – the family has a new room! Such expansions gradually reduce the courtyards to become mere narrow passages and, we were told, caused endless quarrels among residents.

To illustrate the housing conditions, I'll mention a curiosity, an early proposal for the first years of our stay in Xi'an. An idea arose about how to give the tenants better sanitary conditions: When surveying the courtyards, we discovered that where two parcels met in the middle of the block, there was often a small area that in fact had no organized use. If those small areas could be utilized and bound together, this could be a continuous zone, which was named "inner service lane." Here it could be possible to establish infrastructure, as well as public baths and shared laundries for all parcels in the block. The idea was to also shield domestic activities from the tourist gaze. This would be a minimum solution for improved sanitary standards and an inexpensive infrastructure improvement. To illustrate the rapid changes in the district: The idea was soon to become irrelevant since only a few years later the overall standard in the courtyards had improved.

Joint Master courses

In 1991, the long-term cooperation started between architectural faculties of Chinese universities and my own university, the NTNU. The first attempt was cooperation with XJU. Later Tsing Hua University in Beijing, SEU in Nanjing, and XAUAT became our good partners. Teachers and students carried out fieldwork on selected sites at the beginning of the autumn semester. For the rest of the semester, students worked in their respective universities, using the collected data, as they worked on planning and design.

I remember the start of the first course with warm feelings. We were sitting in a classroom, Chinese students in one corner, Norwegians in another, all a bit shy and at the same time full of expectations. Teachers were also a bit tense; it was our first attempt to do this. After a few introductory comments, students were invited to mingle. This meeting between the two groups, with very different backgrounds, but similar interests – namely architecture – was very, very beautiful. Immediately all students started communicating in Norwegian-English and Chinese-English. Spontaneous and impulsive, like kindergarten children, despite their age of 20-25 years.

This first course and all courses thereafter dealt with the development of old housing areas, in cities and villages. Studying old housing areas where people are living and carrying out their daily tasks, trying to adapt modern life to the ancient buildings, and vice versa – adapting old buildings to modern life – implies bringing a range of interesting and crucial issues to the fore.

International doctrinal texts and theories stress two kinds of cultural heritage; the tangible, literally spoken what could be touched, and the intangible, what could not be touched. In more general terms tangible is a notion that could be understood as what could be experienced by our senses, and intangible is what exists as tradition, understanding, skills, belief, experiences, etc., common in the mind of the people of a society. When talking about cultural heritage, one consequently can distinguish between culture and cultural expressions. Culture is then understood as what is common in the mind of people, and what is shared within a specific societal sphere. Cultural expressions are perceptible products of this common understanding – architecture, music, dance, fine art, gardens, handicraft products, spoken language, etc. In everyday conversation, these notions are often mixed up, but for me, the distinction between culture and cultural expression has been helpful for understanding the phenomena. The tangible - the cultural expressions, and the intangible - the culture, are interwoven phenomena. They form a totality of cultural heritage also in the inhabited historic areas, highlighting the continuity of life, and the present as a link between the past and the future. Having the opportunity to follow the development of inhabited historical districts has provided abundant food for reflection.

The historic housing districts in China are often characterized by high density and low housing standards. For Norwegians, who live in low-density areas and typically with high housing standards, confronting these situations is a challenge; a very fortunate and useful one, which raises several questions: how do people maintain their dignity and adapt to

the given conditions, how are the available resources negotiated, and – not least pertinent for our particular field of interests – what is considered a cultural heritage among residents of the district, do they consider the cultural heritage
to be important, and furthermore do they respect this cultural heritage or not.

Investigating existing, inhabited courtyards gave us a good
impression of historic building techniques, architectural
form of the old buildings, social conditions, cultural continuity, (or discontinuity), and development of housing standards.
In the first year the design task was to propose how to upgrade one single courtyard, or to study the potential of combining two or more courtyards side by side.

Inspired by earlier studies of vernacular architecture where
climatic conditions played a crucial role, particularly in marginal housing situations the original focus of the initial
courses was architecture adapted to the climate. Very soon,
by widening the scope of interests, we became more con-
scious of the need to understand the cultural context of the
design. Our first task in DTMD fully demonstrated the complexity of the situation, where the climate was just one of
several factors for finding good design solutions. It soon
became clear that should we limit ourselves to problems
around climate adaptation we would miss a fantastic possibility to learn much more through facing the design challenges with a holistic approach.

Consequently, we sought to develop the joint teaching
programs encompassing three courses in combination:
Architecture and Anthropology, (based on semi-struc-
tured interviews), Documentation and Analyses, (including
measuring historic buildings) and Architectural Design

(based on empirical material from the two other courses). This master course model was employed from year 1995 and onwards, thus developing a solid relationship with our new partner, XAUAT.

With a few exceptions, we carried out the joint master courses with XAUAT, mainly dealing with inhabited urban, housing areas. Traditional housing areas in China have often been located in the urban centers. Therefore they have been removed or threatened to be removed by the rapid urban development. Central areas are attractive for real estate companies, and the low-rise, historic housing districts have, typically, given way for modern, high-density urban blocks. For students it was very educative to be confronted with the contradictions between restoration/conservation, up-grading and modernization of such areas. In Xi'an there were a few historic courtyards when our master classes started the courses, and we ran the courses in DTMD, the Beilin District, and the sign-painter street south of Drum Tower. The two former districts were within areas designated as 'Protection Areas' in the Master Plans, the latter not. There was some bewilderment around what 'Protection Areas' indicated. What should be protected? How? And why? These were challenging questions for anyone, a fine starting point for critical thinking and reflections, which should result in a design. What impressed a visitor like me, grown up in generous dwellings in widely scattered housing areas, was the seemingly dignity of people living here. Out from the rather shabby, physical environment, people came out in the morning, heading for their job. Well dressed, clean and smiling.

05

当不同
文化
相遇
西安
1997—
2003

05

WHEN
CULTURES
MEET
Xi'an
1997-
2003

在异国他乡（即便在自己的国家）工作，意味着要面对不同的文化。我们应该如何处理未知的情况？我们如何在陌生环境中处理自己的角色？西安鼓楼回民区项目的故事，很好地说明了一些可能存在的问题和挑战。

1996年，中国和挪威政府签署了一份环境项目合作合同，其中大多数项目是关于如何解决供水问题，但有一个项目是关于如何在历史住区实现城市的可持续发展。这个项目就是西安鼓楼回民区项目。我所在的挪威科技大学（NTNU）是该项目的挪威方负责单位，西安市城乡建设委员会则被指定为中方对口单位。大清真寺周围11.8公顷＊3的区域被确定为项目区域。这里约有5000多名居民，其中大部分（93%）是回族。中国政府和挪威开发合作署（NORAD）为该项目提供了资金保障。

鉴于复杂的背景，挪威团队面临着许多层面的文化挑战。首先，从挪威这个500万人口的欧洲小国来到中国这个当时拥有12亿多人口的亚洲大国，已经是一种极端体验，我们需要适应并尝试从国家层面了解中国的最基本特征。其次是西安，一个拥有几百万人口的城市（比挪威的总人口还多），以及位于这个城市中心的回族文化。最后，在这种文化中，有三个不同教派及其各自的清真寺。有人建议我们另选地点做项目，"不要选这里的回民区。你们永远不会成功的"，暗示这里的矛盾关系错综复杂，有政府和居民之间的，不同教派之间的，不同家庭之间的，甚至家庭成员之间的。所有这些，对于我们这些在头脑简单的、民主社会主义传统环境中长大的挪威人来说，不是一个容易应付的局面。但我们坚持我们的计划，恪守原有国家层面的挪中协议。

由于有国家层面合作协议的支持，我们的项目得到了西安市政府的高度重视。西安城市改造发展迅速，投资和建设规模巨大，政治决策面临着不少困难。基于这点，我们这个规模相对很小的鼓楼回民区项目所得到的政府关注支持远远超出我们的期望。双方协议的履行委托给了西安市城乡建设委员会，并随即把责任下放给政府的房

＊3　1公顷＝10000平方米。

地产与综合开发处。这样做的理由是，房地产与综合开发处拥有城市土地和相应的资金，可以简化很多程序。多年后，我从一位退休的高级官员口中得知，他对于我们的项目确实存在着不同的意见。不过，我们遇到的官员都表现出支持和鼓励的态度。

一个有趣的问题是中国自20世纪80年代以来市场经济发展所带来的影响。改革开放后，私人房地产开发商和政府房产部门获得了越来越多的权力。私营企业和公共管理部门之间的关系也发生了一定的变化。一开始，双方并不明确谁在建筑事务中能够做出哪些决定，谁在不同层面的事务中拥有最后的发言权。因此，产生了一些困惑和不确定性，但是这些问题都很快得到了解决。

项目任务是一个多方面的城市更新项目，涉及城市设计、基础设施改善、有历史价值院落的保护、嵌入式住宅设计、与居民的沟通、地理信息系统（GIS）的登记和计算机文件编制，以及工作人员的培训。这样一个综合性的项目需要一个多学科的团队，包括建筑学、城市规划、历史、考古、人类学、市政基础工程和计算机科学等方面的专业人员，这些人员都来自挪威科技大学。中方团队的人员主要来自西安市规划院，由建筑师、规划师和文秘人员组成。

作为外来者，面对如此多层面问题的项目，我们需要对工作的内容有一个统筹全面的认识。绝望之际，我和我的同事开发出我们称之为"资源系统法"的方法 * 4，这一方法如同一个生命浮标，成为把我们从混乱头绪中解救出来的分类工具。一天工作下来，会收到来自众多渠道的各种往往自相矛盾的意见，事实证明，这一工具对于区分重要和琐碎、真实和虚假、过去和现在、估算可能性和不可能性都很有助益。其中一个步骤是对资源进行定义和分类。我们就此将所收集到的材料广义地理解为物质、社会文化和人力资源。

社会文化资源的收集既有清晰的结构化资料，也有碎片式的，又或

---

* 4 在1991年的时候，我们就体会到需要对于复杂条件下各种因素开展研究。Bråten and Høyem：The Resource System Method, NTH, Trondheim, 1991.

介于两者之间的资料。所谓结构化，指的是在相对确定的条件下，如宗教信仰、政治权力模式、国家与国际标准和协议，等等。所谓碎片式，指的是更加多变的伦理观念或模式；也许是一种受内部矛盾或外部影响对现状的质疑，或受到新思想、新技术和新知识等的影响。

整个鼓楼回民区划归四个居委会管理，区域内有十座清真寺。回民区中又有不同的教坊，即围绕一座清真寺组成一个小的穆斯林社区。*5走在大街上，不同教坊间的边界并不明显。大清真寺的阿訇在市政协中代表回坊所有穆斯林。*6为征求他的意见，我们经常见面。在我们工作的30年间，上述社区组织模式没有太多变化。

因为旅游业的蓬勃发展，鼓楼回民区很多餐馆的菜单渐渐发生了变化，例如，允许给顾客提供酒水饮品。更为显著的是市场经济和改革开放政策带来的变化。旅游业和地方商贸活动受到鼓励。1985年，我第一次到访该片区时，只有几个小贩提供商品，几乎没有临街的店面商铺。后来，市场和商店发展起来，随着旅游业升级，当地的商业也蓬勃发展起来。大清真寺吸引着朝圣者，特别是在斋月期间，有来自国内外的朝圣者。历史上西安是丝绸之路的起点，贸易于回族一直占有重要地位。人们可以认为，现代旅游业（包括朝圣）和今天的贸易活动是古老传统的延伸，从而也增强了地方认同。

一般来说，分类时涉及与人群有关的内容都归类于人力资源材料。就西安鼓楼回民区来说，一系列群体和个人都属于这一类别。换言之，他们是项目的参与者，对我们的整个研究过程产生很大影响，包括这里的居民、专业工作者、政府人员、当地商贩、房地产开发商、宗教领袖。人们凭借他们的地位和个人的天赋、能力和个性发挥着他们的作用。我们这样的长期项目提供了一个更好地认识和了

---

*5　Dong Wei: An Ethnic Housing in Transition. Chinese Muslim housing Architecture in the framework of resource management and identity of place. NTH, 1995.

*6　译者注：即西安化觉巷清真大寺伊玛目阿訇马良骥（1931—2018），知名伊斯兰学者与爱国宗教人士，曾任全国政协委员、中国伊斯兰教协会副会长、陕西省伊斯兰教协会会长及名誉会长、西安市伊斯兰教协会会长等职。

解这些个体以及他们与周边人群关系的机会。我们与鼓楼回民区的联系已有10—12年。尽管这种长期视角有好处，但也有一些因素干扰限制了我们对情况的理解。例如，政府和专业机构中十几年来工作人员的变化；再如，尽管我们试图保持开放、中立的态度，但某些文化差异带来的困难还是没有办法克服。当然，还有语言问题，除了一位人类学学者外，挪威团队中没有人会讲中文，大多数对话都需要翻译。

我们一方面与院子里的家庭定期接触，另一方面与当地政府和宗教机构的官方代表联系，如与街道居委会成员和清真寺领导交谈，了解居民的情况。我们花了一些时间搞清楚当地三个教派及其与清真寺之间的关系，以及大清真寺阿訇的作用。马阿訇是我们的主要联系人，他非常睿智、和蔼和友好，他的权威身份也给我们提供了很多有用的意见和建议。阿訇代表整个回族社区与政府沟通（我们没有获得任何关于这方面互动的第一手资料）。对我们来说，这是当地社区社会网络强大最明显的表现。和我们交谈过的每个人都表示愿意留在回民区生活，明显表达出对自己居住地方的归属感和依恋。

我这里用一个例子说明研究过程中了解当地的社会关系为何至关重要。我们对传统院落的保护共提出了四个建议方案，其中一个方案未能实施。为什么会这样呢？

项目地块宽10.5米，进深54米。该地块原为两兄弟所有。除最内侧区域外，该地块被纵向分割。外侧面向街道部分为共有，并分割为一左一右两部分。这个大家庭中的十户人家都住在这里，其中有一位单身老祖母，父母辈成员16人，子孙辈20岁左右的有6人，2—8岁的有7人。对于保护方案，这个大家庭中有些人接受欢迎，有些人持怀疑态度，但愿意接受一定改变，有些人则反对任何干预和变化。因此谈判非常复杂。经过一年多的对话，居民们似乎对改造计划达成了一致，并指定一位家中的中年男性成员代表他们参与进一步协商环节。离最后签署协议的时间越来越近了，所有人集合在一处，准备正式签合同。但突然间，那位老太太拒绝签字，她恰好是其中一个老房主的妻子。没办法，保护项目遂因此搁浅。

鼓楼回民区也住着一些汉族家庭，但回族人数要多得多，在当地生活中占据主导位置。汉族居民和回民和平共处，在街上经营着各自的生意。

图 .05-1
与保护院落家庭
成员讨论方案
（西安，1998年）

Figure.05-1
Meeting in
the Muslim
District
Xi'an (1998)

图 .05-2
挪威时任教育
部长访问
大清真寺
（西安，1999年）

Figure.05-2
Official
visit: the
Norwegian
Minister of
Education
and Research
visiting the
Great Mosque
Xi'an (1999)

图 .05-3
挪威电视片制作
团队在保护院落
进行采访
（西安，2004年）

Figure.05-3
Film producers
from the
Norwegian
Broadcasting
in a courtyard
Xi'an (2004)

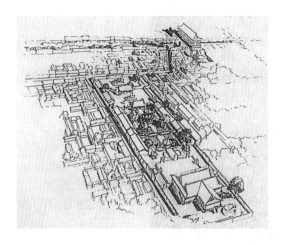

图.05-4
鼓楼大清真寺周
边环境概念草图
(图源：陕西省城乡规划设计
研究院，1997年)

Figure.05-4
Sketch
illustrates
the original
concept of the
surroundings
of Drum Tower
and the Great
Mosque.
(Photo credit: Shaanxi
Provincial Urban and Rural
Planning Institute, 1997:20)

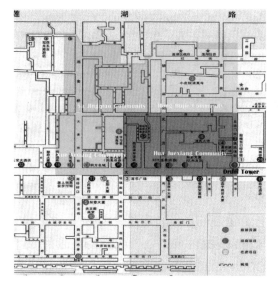

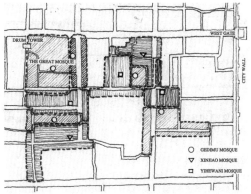

图.05-5
回坊社区平面
组织关系图
(图源：作者自绘教学课件)

Figure.05-5
The
organization
patterns of Hui
community in
Xi'an's DTMD.

图.05-6
在化觉巷购物
的游客
（西安，2002年）

Figure.05-6
Tourists
visiting the
Huajue Lane
in DTMD
Xi'an (2002)

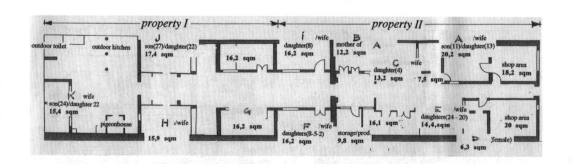

property I								property II	
outdoor toilet	outdoor kitchen	son(27)/daughter(22) 17,4 sqm	16,2 sqm	i /wife daughter(8) 16,2 sqm	B A mother of 12,2 sqm	C daughter(4) 13,2 sqm	/wife 7,5 sqm	A /wife son(11)/daughter(13) 20,2 sqm	shop area 18,2 sqm
K wife son(24)/daughter 22 15,4 sqm	pigeonhouse	H J/wife 15,9 sqm	16,2 sqm	G daughters(8-5-2) 16,2 sqm	F /wife storage/prod. 9,8 sqm	16,1 sqm		E /wife daughters(24 - 20) 14,4,sqm	shop area 20 sqm D (female) 6,3 sqm

图.05-7
相关院落人口分
布平面图（A-K
代表每户的户主）
（图源：西安鼓楼回民区西羊
市街4号院挪威团队研究报告，
1999年）

Figure.05-7
The plan
shows the
distribution
of people in
the courtyard
described

below. A-K
indicates the
head of each
household.
(Photo credit: Research on
future potential - Xiyangshi
Street No. 4 Report, Xi'an,
1999)

图 .05-8
从面向街道打开的店门可以看到这个家庭如何在15平方米的 房间里生活，养育孙女，同时经营着裁缝生意（西安，2009年）

Figure.05-8
Through a door facing the street one could peep into a 15 sqm room where a family was living, bringing up grandchildren, and running a tailoring business Xi'an (2009)

图 .05-9
另一种将家庭和生意结合的方式：在后院制作食品，在面向街道的店面销售（西安，1998年）

Figure.05-9
Another way of combining home and business: Living and producing food for sale in the courtyard behind and selling the products from a small shop facing the street Xi'an (1998)

Working in a foreign country (and even in one's own country) means confronting many-sided cultural conditions. How should we deal with the unknown aspects of the situation? How do we handle our own role in the unfamiliar context? The story of the DTMD project in Xi'an illustrates what may be at stake.

In 1996, the Chinese and the Norwegian government signed a contract of cooperation on environmental projects. Most of these were projects dealing with water supply problems, but one was a project on sustainable urban development and development in a historic housing district. This project was DTMD in Xi'an. My university, the NTNU, was given the responsibility to carry out the project on the Norwegian side. Xi'an Urban and Rural Construction Commission were appointed to be the counterpart on the Chinese side. An area of 11.8 hectares, surrounding the Great Mosque, was defined as the project area. This area had approx. 5,000 residents, most of which (93 percent) belonged to the Hui ethnic group. Financing was guaranteed by the Chinese government and by the Norwegian Agency for Development Cooperation (NORAD).

Granted the complex context, there were many levels of cultural challenges for the Norwegian team. To start with, going from Norway, a small European country of 5 million inhabitants, to China, a vast Asian country of 1.2 billion inhabitants, was initially an extreme experience. We needed to adapt and try to understand the most basic features of China on nation level. Then there was Xi'an, a city of several million people (more than the entire population of Norway), and furthermore the Hui Muslim culture in the middle of this city. And finally there was the three distinct sects with their separate mosques inside this Hui culture. Some people advised us to find another site for our project. "Leave the site. You will never succeed." one person said. Then, hinting at contradictions between the different sects, between different families, and even between individual family members, and, of course, differing views on the development among authorities, residents, and other stakeholders. Although it is not an easy situation to cope with for us Norwegians, dewy-eyed, and raised in a social-democratic tradition, we stuck to our plan, loyal to the original state-level Sino-Norwegian agreement.

Because of the backing of this state-to-state cooperation agreement, our project was given high priority by the city government. Considering the rapid urban renewal development of Xi'an during these years which meant difficult political decisions, heavy investments, and comprehensive large-scale constructions – and considering the relatively small scale of the DTMD project – we gained far more attention from the political system than could be expected. Fulfillment of the agreements was delegated to Xi'an Urban and Rural Construction Commission, who thereupon delegated the responsibility to the governmental Real Estate Division. The argument for this was that the Real Estate Division possessed urban land and had the financial muscles to ease the processes. I heard from a retired, high-level politician many years later, that there were indeed contradicting views on our project, also behind closed doors. However, the politicians we met were both supportive and constructive.

One interesting issue was the effect of the development of a market economy in China since the 1980s. During these years private and public real estate actors gained more and more power. The change of roles in the relationship between the private and public sectors became a factor. In the beginning, the parties were not sure who made which decisions in building affairs. Who had the last say in matters on different levels? And thus the situation could have been confusing and unpredictable. But rather soon these matters were untangled.

The task was a manifold, urban renewal project, involving urban design, infrastructure improvement, conservation of historic, valuable courtyards, infill housing design, communication with residents, registration, and computerized documentation on Geographical Information System (GIS), and staff training. This comprehensive project needed a multidisciplinary team, consisting of professionals in architecture, urban planning, history, archaeology, anthropology, infrastructure engineering, and computer science, all recruited from NTNU. The team on the Chinese side was manned mainly by the City Planning Bureau, and consisted of architects, planners, and secretaries.

It is easy to understand that when coming from the outside and faced with such a multifaceted field of problems, it was demanding to comprehend the totality of what we were working with. In despair, my colleague and I developed what we named the Resource System Method, * 2 a kind of lifebuoy, a sorting tool to survive the chaos in our minds and hearts. After a day full of various, often contradictory input from numerous sources, this tool turned out helpful for separating important from trivial, true from fake, past from present, and possible from impossible. One step on the way was to define and categorize resources. In this context, we understood resources broadly, as material, sociocultural and human resources.

---

*2　Bråten and Høyem, The Resource System Method, NTH, Trondheim 1991.

Sociocultural resources can be found somewhere in between the structured and the fragmented. By structured I mean relatively fixed conditions, like religious faith, political power patterns, national and international standards and agreements, etc. By fragmented I mean patterns, ethical conceptions, or models that are more flexible and might even be in the process of changing; maybe there is a questioning of the status quo, initiated by internal contradictions or external influences, affected by new ideas, new technology, new knowledge, and so on.

The governmental organization of the DTMD was based on a subdivision of four Neighborhood Committees. In the whole area, there were 10 different mosques, which belonged to three different sects. They were organized in the so-called Jiao Fangs, a community with a mosque surrounded by members of a particular sect. * 3 Walking the streets the borders between the different Jiao Fangs were not visible. The leader of all the Huis was the Imam of the Great Mosque, who in the city board represented all Muslims. Seeking advice from him, we often met and could observe that the pattern mentioned above did not change much during the period of 30 years we worked there.

Other religion-related issues were slowly under attack, not least because of the thriving tourism. For instance, some restaurants, even in the DTMD, did introduce alcohol to their customers. Little by little menus in many restaurants changed. More conspicuous was the influence of political changes, notably the introduction of the market economy

---

*3   Dong Wei: An Ethnic Housing in Transition. Chinese Muslim Housing Architecture in the Framework of Resource Management and Identity of Place. NTH, 1995.

and the Opening-up policy. This encouraged tourism and local trade, also here in DTMD. The first time I visited the district, in 1985, there were just a few hawkers offering their
merchandise and hardly any shops facing the small streets.
Later markets and shops developed, and along with the escalating tourism, local trade flourished. The Great Mosque
started to attract pilgrims, notably during Ramadam, not only from other provinces but also from other countries. Historically Xi'an was the terminal of the Silk Road, and among the
Huis here trade has always been very important. One could
maintain that modern tourism (including pilgrimage) and
today's trade is an extension of old traditions and thus en-
hances local identity.

Generally speaking, in processes where people are involved,
they will fall within the category above named human resources. In the case of the DTMD, a range of groups and individual persons fell within this category, in other words, were
actors in the projects and heavily affected the processes we
studied: residents, professionals, authorities, local business-
men and -women, real estate developers, religious leaders.
Persons played their role by virtue of their position and by
their individual talent, capacity, and personality. Long-term
projects like ours offer a chance to better get to know and
understand individuals and the relationships between them.
Our contact with the district developed over 10-12 years. Despite the benefits of this long perspective, there were also
conditions that disturbed and limited our comprehension of
the situation. To mention a few: The change of staff on several levels of the political and professional systems; discussions and negotiations which were relevant for the project
but took place in religious and political fora where we had
no access and from which we only got second-hand information; cultural differences which were hard to navigate de-

spite our attempts to be open-minded and neutral. And then language problems, of course – with the exception of one of the anthropologists, no one in the Norwegian team could speak Chinese and needed a translator in most dialogues.

We learned to know the residents partly through scheduled contact with families in the courtyards and partly through speaking with local, official representatives of the political and religious systems, like street committee members and mosque leaders. It took a while to understand the relationship between the three sects and their mosques, and the role of the Imam of the Great Mosque. He was our main contact; a wise, kind, diplomatic, and friendly man with the necessary authority to properly advise us. He represented the Hui society as a whole and negotiated with the city authorities on disputes between the locals and the authorities. This was – for us – the most distinct public manifestation of how strong the local social networks were. Everyone we talked to expressed their will to stay and live in the district. The attachment to the place was substantial and evident.

As an example of how important it was to understand these local, social relationships, I will mention one case. One of our four suggested conservations of a traditional courtyard was not carried out. Why not?

The parcel was 10.5 meters wide and 54 meters deep. The site was originally owned by two brothers. Except for the innermost parcel, the property had been divided lengthwise. The outer parcel, facing the street, was shared and subdivided into a left and a right part. Altogether there were 10 households of the same family living there, consisting of one single, old lady, 16 of the parent generation, 6 children in the twenties, and 7 children 2-8 years old. Some welcomed the conservation plan, some were skeptical but open to a change, and some were against any intervention. Obviously, these negotiations turned out to be complicated. After more than a year of dialogues, the residents seemed to agree on the plan and appointed a middle-aged man to represent them in the further process. We were getting closer to the signing of the final contract, and the adults were collected for the formal signing. But then suddenly the old lady, who happened to be the wife of one of the original owners, refused to sign. No way. And that was the end of that conservation project.

There were some Han families in the district, but the Hui people were far more numerous and dominated the local life. The Han and the Hui residents lived peacefully together, running their businesses side by side in the streets.

06

生活在
文化
遗产中
西安
1997—
2003

06

LIVING
IN A
CULTURAL
HERITAGE
Xi'an
1997-
2003

在全球范围内，传统住宅逐渐被视为一个国家文化遗产的重要组成部分。过去几十年来，在快速发展的城市中，有人居住的历史街区一直承受着巨大的压力，甚至面临着被全面拆除的危险。为什么会这样？这是很多因素的共同作用，我想略谈一下在中国的案例中比较典型的几个因素。

城市的快速发展需要占用更多的土地。这些土地可以来自现有城区的外围，也可以在城市中心地带，因为那里原有的低层建筑可以代之以高层建筑。后一种情形无疑对寻求巨额利润的房地产开发商充满诱惑。而在历史居住区，很多家庭住在拥挤的四合院中，住房标准急需提升，而这就意味着那里有着对更大住房的需求。而随着时间推移，由于资金匮乏，有人居住的历史街区多疏于维护，处于破败之中。

一般来说，农村的土地压力比城市更小，情况相对更简单。在中国，不断扩展中的大城市所面临的问题最为严重。大城市中大多传统住区已被抹去，保护其剩余者免于衰败和破坏变得至关重要。不过，公众的保护意识也在提升。这一方面，源于人们对文化遗产本身的兴趣，另一方面，因为这些为数不多、硕果仅存的街区成了旅游景点。

在西安，鼓楼回民区及大清真寺是仅次于秦始皇兵马俑的重要旅游景点。作为这个城市中仅存不多的历史街区之一，这里的建筑和街道生活对游客充满吸引力。我们的项目包括四个传统院落的保护。在1997年开始修复工作时，传统院落已经所剩无几。遗憾的是，我们最终只修复了其中三座院落。*7如第五章所述，第四个院子中的住家就是否参与保护项目不能达成一致。尽管通过挪威外交部发展合作署（NORAD）获得了项目资金，但四号院的住家对整个项目疑虑重重，最终选择不参与此事。

---

*7　译者注：中国挪威联合团队最终完成保护与康复性更新的三座传统院落分别为：西安北院门144号院（高岳嵩故居）、化觉巷232号院（原125号）、西羊市77号院。

与此同时，建筑方面也存在着其他问题。该片区1989年时的住房条件很差，人均住房面积为2.7平方米。不仅居住面积小，卫生间、厨房、供水设施亦阙如。原本一户人家的院子，现在通常都有很多住户，这样的密集程度使宽敞的内院变成狭窄的过道。房屋的技术质量水准也不尽如人意。调查显示，能够修复和保存的传统四合院屈指可数。在与各方人员协商合作下，我们确定了其中四个院落，将其设定为保护工作的对象。

我们选择了四个院落中最小的一座作为起步项目，其原因有三：首先，只有一户人家住那儿，这样更容易沟通。其次，它是面积最小的房产，因此相对最简单。最后则是房屋的技术质量状况看上去不错。当然，最后这个预判是错误的。当工程开始，揭去表层油漆后，我们发现大部分木构件已经朽烂，必须更换。我们所有的操作严格按照教科书的要求进行，高标准、专业化地完成了修复。因为修复结果甚佳，以及住户和专业团队之间建立的融洽关系，这个院落后来获得了联合国教科文组织的荣誉保护奖。*8院子主人在整个项目中发挥了重要的作用。在正式同意参与前，他仔细调查了项目的背景。项目开始之初，也曾经历一次危机，房主的儿子想终止与我们的合作，因为他原本计划拆掉整个院子，建一栋全新的小楼。政府和他父亲否决了他的想法。在整个保护工程中，房主是一位忠实的合作者，还为好奇的左邻右舍介绍整个项目过程。*9最后，当修复工作结束时，这座院子的成功也影响了当地居民和政府官员对整个项目的看法，连最初持反对意见的儿子也上前对我们的努力表达了谢意。

在处理院落修复问题时，大家一致认为我们应遵从国际宪章文本和保护标准。在这个过程中，参与各方对文本又有各种有趣的不同理解和解读，这里仅择其要点略予简述。

---

*8　译者注：化觉巷232号院（原125号）的保护修复在2002年荣获联合国教科文组织亚太区文化遗产保护奖荣誉保护奖（Honorable Mention, UNESCO Asia-Pacific Heritage Awards for Culture Heritage Conservation），参　见：https://bangkok.unesco.org/sites/default/files/assets/article/Asia-Pacific%20Heritage%20Awards/files/2002-winners.pdf.

*9　译者注：化觉巷232号院（原125号）肇建于清乾隆年间，为安氏祖宅，作者文中所述房主即回族书法家、西安理工大学退休教授安守信。

"真实性"是一个复杂的概念，在保护工作中发挥着重要作用。在合作过程中我们发现，对挪威团队来说，文物的现状、原始技术和原始材料的使用，以及对"真相"的探寻最为重要。而对中国同行来说，其原初形式的完美理想化概念最重要。这种差异在我们阐述勘察和测绘图纸时愈加明显。挪方希望将建筑中所有不规则的细节都纳入图纸中。而中方则不然，尽管所有数据都是基于现场测绘，中方团队实际上在图纸中描绘了对具体建筑的理想化概念，抹去了现状中的不规则之处。

在涉及材料和技术的真实性方面，工程严格按照国际宪章准则实施，与其他不太科学严谨的工程相比，这是比较少见的。施工队伍由在很多重要古迹项目中有实践经验的聪明匠人组成。他们中的一些人是农民，在自家农田收割的季节不得不离开施工现场。这对时间的安排造成了些许干扰，但这些干扰影响甚微，因为工人们总是在农活结束后立即返回工地。

政府通常尽量少地干预回民区的地方事务。这里的基础设施标准落后于西安市其他居住区的正常标准。在项目初期，按照总体规划要求需要拓宽几条街道，理由是交通和地下基础设施的现代化需要更多空间，而这似乎与总体规划的另一项规定相抵牾，即该片区被归类为"保护区"，拓宽会影响街道的尺度和面向街道的建筑，换句话说，它会影响到那些被明确认为有保护价值的要素。尽管有总体规划的规定，但保持街道现有宽度的观点在随后几年中得到推重。该片区因此保持了其特殊的街道氛围，这点也受到当地居民、游客和西安其他地区居民的欢迎。虽然交通密度依然很高，但机动车交通已经明显减少。

还有对建筑高度的规定，目的是保持鼓楼和大清真寺这两处主要古迹在城市景观中的突出地位。遗憾的是，很多建筑违反了这一规定，具体原因我们无从得知。也许我们可以这样安慰自己：与市中心其他地方相比，鼓楼回民区的发展还是相当温和的。希望我们的项目能够为该片区的保护提供一些动力，阻止高层建筑的建设，保持该片区作为一个特殊街区的特点，并尊重其作为保护区的地位。

基础设施问题与现有建筑物和院落直接相关。供水、排污管道以及电力线必须入地。国家对这些地下管网的深度和宽度都有规定，但是如果完全遵守这些规定，周边建筑的地基就会受到影响，进而危及建筑本身。唯一的解决办法，就是像在历史环境中经常出现的情况那样，为这些片区找到替代性的、适宜的技术性解决方案，并出台相应的特殊规定。经过多年的谈判，政府最终接受了在国家标准基础上加以调整的建议。于是，历史街区的公共空间得到了保护，数百人居住的院子有了供水和排污系统。这可能是该项目对当地居民最重要的贡献。当管道、电缆和由坚实的天然石板铺就的崭新街面到位后，项目团队获得了广泛的认可。

保护的一个基本先决条件是，所有生活在该片区的居民都可以留下来，这个先决条件一直得到项目组的推重。当时的总体问题是：如何能在尊重该片区历史特征的同时，提升总体居住水平，实现其现代化？这种总体住房水平的提升应该是怎样的？新建的建筑是应该采用所谓"历史风貌样式"，还是应适应街区内大量新建建筑的主导风格——即董卫在一篇文章中所谈的"新乡土"？是否应制定一个关于形式的指导手册？如果这样做，这一指导手册的规定应当强制实施抑或只是一种建议？

我们选取了两种策略：一个是保护四个指定的传统院落，另一个是实施一些嵌入式设计项目，以呈显该片区在规范下的可能性和边界限度。这两种策略，都应为那些希望改善住房条件的家庭提供一些灵感和参考。项目办公室应作为服务和信息中心面向公众。不难看出，整个项目过程会遇到很多妥协，但三个传统院落的保护是非常成功的。项目办公室成立后，在作为公众信息咨询中心方面所起的作用有限，但在其他方面则发挥了应有的作用。经多次尝试，我们只实施了一个嵌入式住宅项目。挪威团队的设计没有被采纳，最终采用的是中方团队的设计方案，但方案随后被开发商修改，在原设计上又加了一层。因为没有一个官方批准的保护规划可供依循，这个项目最终是失控的。

我们的目的是教科书式地遵循保护原则，对四个历史院落进行修缮。由于第五章中所述及的原因，我们不得不放弃了其中一个院落。

其余三座院子则为应该如何去修缮提供了良好的示范：只维修需要维修的东西；只替换需要替换的东西；如果替换确有必要，则应使用原来的材料和技术。

为什么这个城市中心区的黄金地段没有被一些有能力的房地产开发商买下来用于营利式开发？其实有过几次尝试。政府曾让一位开发商负责整个片区的现代化建设，但是这个任务显然超出了他的能力。除了屈指可数的几个例外，即使是买下两个相邻的院子，将其作为一个单元开发，也被证明很难实现。唯一可行的方案似乎是在一个地块范围内扩大商铺面积和新的住房面积。我们最近看到，市场力量在慢慢成形，正改变着这一历史片区的特征。

图 .06-1
化觉巷大清真寺
周边住区鸟瞰
（西安，2002年）

Figure.06-1
Aerial
view of the
neighborhood
around the
Great Mosque
Xi'an (2002)

图 .06-3
洗衣设施
（西安，1995年）

Figure.06-3
Laundry
facilities
Xi'an (1995)

图 .06-2
公共自来水
（西安，1995年）

Figure.06-2
Water supply
Xi'an (1995)

06 生活在文化遗产中·西安·1997—2003

图 .06-4
保护维修前的化
觉巷原125号院
（西安，1995年）

Figure.06-4
Courtyard No.
125 on Huajue
Lane before
conservation
Xi'an (1995)

图 .06-5
保护维修中的屋
架与糟朽梁木
（西安，1999年）

Figure.06-5
During
conservation
Xi'an (1999)

图 .06-6
保护维修后的化
觉巷原125号院
（西安，2005年）

Figure.06-6
After
conservation
Xi'an (2005)

图 .06-7
保护完好的砖雕
（西安，1996年）

Figure.06-7
Well-
preserved
stone carving
Xi'an (1996)

图 .06-8
由维修后院内
看铺面
（西安，2005年）

Figure.06-8
View from the
backyard
Xi'an (2005)

图 .06-9
后院新加卫生间
改造前后
（西安，
1996/2002年）

Figure.06-9
New bathroom
in the
backyard after
conservation
Xi'an
(1996/2002)

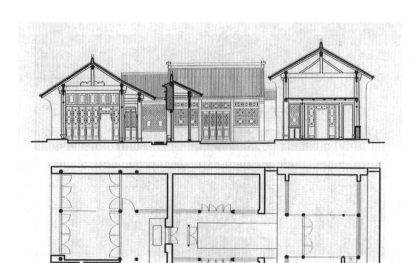

图 .06-10
化觉巷原125号
院落测绘图

（西安鼓楼回民区挪威团队工
作报告，1999年）

Figure.06-10
Survey
drawings

(Photo credit: The DTMD
Report - No. 125 Huajue
Lane, Xi'an, 1999)

图 .06-11
联合国教科文
组织颁发的奖
牌、奖状
（西安，2007年）

Figure.06-11
UNESCO
reward
Xi'an (2007)

图 .06-12
缺少"真实性"
的改造例子
（西安，2012年）

Figure.06-12
Lack of
authenticity –
examples from
Xi'an (2012)

图 .06-13
新的铺面显示道
路拓宽已非必要
（西安，2018年）

Figure.06-13
New pavement gives an indication that the street widening will not take place Xi'an (2018)

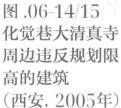

图 .06-14/15
化觉巷大清真寺
周边违反规划限
高的建筑
（西安，2005年）

Figure.06-14/15
Breaking height regulation rules Xi'an (2005)

图 .06-16
嵌入式住宅项目
改造前、后状况
及挪威团队设计
方案模型
（西安，2004/
2007/2004年）

Figure.06-16
Original
situation,
realized project
in Xi'an, and
NTNU project,
Trondheim
Xi'an (2004/
2007/2004)

图 .06-17
从鼓楼上看化觉
巷清真大寺及回
民区
（西安，2019年）

Figure.06-17
View from the
Drum Tower
on the Great
Mosque and

surrounding
area
Xi'an (2019)

Globally, there has been increased interest in looking at traditional housing as an important part of a nation's cultural
heritage. In the last decades, the inhabited, historic districts
of fast-growing cities have been under high pressure, and in
danger of total demolition. Why is this? Many factors work
together, and I will mention a few that have been typical for
Chinese cases.

When there is rapid urban growth, there is a need for more
land. This land is found on the periphery of the existing, old
cities, but also in the city center where low-rise buildings
can be replaced by high-rise blocks. This last scenario, of
course, is very tempting for real estate businesspeople looking for profits. With many families living in congested courtyards in the historic residential areas, the nationwide housing standard upgrade means a need for more housing. Over
time, because of a lack of funds, scant maintenance of inhabited, historic districts resulted in decay.

Normally, the situation is easier in the countryside villages,
where the territorial pressure is softer. In China, the situation has been gravest in the growing, big cities. Most of the
traditional, housing districts have been erased, and it is critical to protect the remaining ones against dilapidation and
destruction. Partly because of interest in cultural heritage in
itself, and partly because these few, remaining districts are
tourist attractions, public awareness around this problem is
on the rise.

Only second to the Terracotta Warriors, DTMD, with its
Great Mosque, is the main tourist attraction in Xi'an. As it is
one of the very few remaining historic districts in the city,
the appeal is to a large extent the architecture and street
life. Our project encompassed the conservation of four tra-

ditional courtyards. (Very few were left when we started the restoration in 1997). Unfortunately, we managed to restore only three of them. As described in chapter 5, the families in the fourth courtyard could not agree to take part. Despite the fact that funding for the project was secured - through the Norwegian Agency for Development Cooperation (NORAD) - the families in the fourth courtyard were too skeptical of the whole project, and consequently chose not to participate.

On the building end of things, there were other problems. In 1989, the housing standard in the district was very low. The average housing area was 2.7 sqm per person. Not only was a shortage of dwelling areas a problem, there was also a deficit in kitchen facilities, water supply, and bathrooms. The courtyards that originally housed one family were now typically home to many families, and the densification changed a spacious yard to a narrow corridor. Technical standards were also poor. A survey showed that there were only a few traditional courtyards that could be restored and saved. In agreement and cooperation with the actors involved, we identified four of these as important, local cultural heritage, and appointed them to be conserved.

To start with, we selected the smallest of the four properties, for three reasons: Firstly, only one family lived there – and this would make communication easier. Secondly, it was the smallest property and thus least complex. And, finally, because it looked to be in a healthy, technical condition. The last assumption was wrong. As soon as a coat of paint was removed, we discovered that a majority of the wooden elements were rotten and had to be replaced. It was all done according to the textbooks, very professionally and at high standards. Because of the results, and because of the good relationship between the tenants and the professional team, this courtyard received a UNESCO honorary mention. The owner of the courtyard played an important role in the whole project. He carefully investigated the background of the project before consenting to take part. There was a crisis at the beginning, when his son interrupted and wanted to stop everything. He had planned to demolish the courtyard and replace it with a housing block. His protest was turned down by a combination of governmental power and his father's paternal influence. During the conservation works the owner was a loyal co-player, also informing curious neighbors about what was going on. Finally, when the work was finished, the successful work with this courtyard impacted public opinion about the project, among local residents, and among governmental officials. Even the objecting son came around and expressed gratefulness for our efforts.

There was an agreement that we should refer to international doctrinal texts and standards of conservation when working with the courtyards. During the process, the reading of those texts, by the various actors involved, led to some interesting variations in interpretation. I will mention a couple of them.

Authenticity is a complex notion and plays an important role in conservation matters. During our collaboration, it was revealed that the present state of the object, the original techniques and use of original materials, a search for "the truth," was superior for the Norwegian team, while the idealized conception of the original form was the highest priority for our Chinese counterpart. This became clear when we elaborated on surveys and survey drawings. The Norwegians wanted all irregularities to be included in the drawings. But-even though they were based on the surveys

-the actual drawings produced by a Chinese team pictured the ideal conception of the buildings and rubbed out the existing irregularities.

When it came to the authenticity of materials and techniques, the project was carried out faithful to the international doctrines, which is quite rare compared to less scientific projects elsewhere in Xi'an. Construction teams were made up of clever craftsmen who had honed their skills in the conservation of important monuments. Some of them were farmers and had to leave the construction site when harvesting took place in their own farmland. This caused slight disturbances in the time schedules, but these were almost negligible since the workers always returned to our work site immediately after finishing their farm work.

The authorities had paid much attention to interfering minimally in local affairs. The result was that the infrastructure standard lacked behind the normal standard in other housing districts in Xi'an. The valid Master Plan at the beginning of the project period prescribed a widening of several streets, arguing that both traffic and modernization of the underground infrastructure needed more space. This seemed to contradict another regulation of the Master Plan, namely that the district was categorized as a "Protection Area." A widening would affect both the scale of the streets and the buildings facing those streets. In other words, it would affect elements that had been explicitly deemed protection worthy, and houses had to be removed if streets were widened. The argument for maintaining the existent width of the streets was–despite the Master Plan regulations–respected in the subsequent years. As a consequence, the district maintained a special atmosphere that is welcomed by local residents, tourists, and the rest of Xi'an's residents. Traffic density is still high, but car traffic, in particular, has decreased.

There were also building height regulations, put in place to allow the two main monuments – the Drum Tower and the Great Mosque – to maintain a prominent presence in the urban landscape. Unfortunately, there are many buildings breaking this rule, for reasons we can only speculate about. We can, maybe, console ourselves by maintaining that development in DTMD, compared to everywhere else in the city center, is fairly modest. Hopefully, our project has contributed by offering some impulses to the area, preventing the building of some high-rise buildings, recognizing the character of the district as a special urban quarter, and honoring its classification as a Protection Area.

The infrastructure problems were directly linked to the existing buildings and courtyards. Water supply pipes, sewage tubes, and electricity cables have to be laid in trenches. There were national regulations for how deep and how wide those trenches should be. If these regulations were to be literally obeyed, the foundations of the surrounding buildings would have been affected, thereupon endangering the buildings themselves. The only solution, as often is the case in historical environments, was to find alternative but adequate technical solutions for the districts, and accordingly introduce special rules. After years of negotiations, the authorities finally accepted the suggested deviances from national norms. The public space in the historic area was then saved, and hundreds of people got water and sewage for their courtyards. It was probably the most important contribution to the local residents. When pipelines, cables, and new street covers of solid, natural flagstones were put in place, the team gained popularity.

A basic precondition was that all people living in the district should be able to stay, and this precondition was always respected by the project team. The overall problems were then: How could we modernize and improve the general housing standard, while we also respected the historic character of the district? And what did this improvement of the general housing standard look like? Should new buildings be built in a so-called "historic style," or should they adapt to the dominant style of the great numbers of new buildings in the district – the "new vernacular," as Dong Wei called it in his dissertation? Should a form manual be elaborated? And if so, should it be considered a recommendation or should adherence to the manual be mandatory?

We chose two strategies: One was the conservation of the four appointed traditional courtyards, and the other was to carry out a few infill projects, in order to show possibilities and limitations within the regulations of the district. Both strategies should serve as inspiration and information to the families who wanted to improve their housing conditions. The project office should serve as a service and information center for the public. It is easy to see how this process would involve many compromises, but the conservation of three traditional courtyards was very successful. The project office was established and functioned only to a limited degree as an information and consultancy center, while otherwise, the office functioned as intended. Of many attempts, only one in-fill project was implemented. The design of the Norwegian team was turned down. The design produced by the Chinese colleagues was accepted but was then altered by the developer who added a floor to the original design. In the lack of an approved conservation plan, this project became impossible to control.

The intention was that the conservation of the four historic courtyards should be carried out according to textbook principles. One courtyard had to be abandoned for reasons described in chapter 5. The other three became good examples of how things ought to be done: That is, repair only what needs to be repaired; replace only what needs to be replaced; if replacing is necessary, use original materials and original techniques.

Then the next question is why this shining diamond of an urban center area was not purchased by some solvent real estate manager for profitable development. There were actually several attempts. The government gave a Muslim developer the task of modernizing the whole district. He too was unable to get a grip on the task. There were a couple of exceptions, but even buying two neighboring courtyards in order to develop them as one unit turned out to be impossible. One achievable option seemed to be an expansion of the shop area and new housing volumes within the limits of one parcel. Lately, we have seen that market forces slowly and gradually establish themselves and change the character of the district.

07

户外生活
1985—
2016

07

OUTDOOR
LIFE
1985-
2016

在中国整个工作期间，观察公共领域——在庭院、公园、街巷乃至人行道上的生活日常，于我是一件乐事。对于一个外国人来说，这些活动富有异国情调和吸引力，而且这些活动并没有随着时间推移发生太大变化，它们似乎是文化的真正体现，不受旅游和商业利益影响，历久常存。我的一位朋友对此曾感慨道：观察这些户外生活令人振奋，它重新唤起了我对人类的信心。

当我们观察中国正在经历的快速而巨大的变化时，值得注意的是，有些东西似乎变得更加持久。虽然它们也在变化，也在发展，但其速度与其他事物不同。在这一章，我将用比其他章节更多的图片展示这些并未很快改变的事物。这种持久性能告诉我们什么？人们为什么需要一些比较持久的东西？也许这些东西对人有着更深层次的吸引力，更加难以舍弃？那么，我们所谓"更深层次的吸引力"是指什么？比起那些日复一日的行为，这种吸引力是否在更多的层次上触及了我们的存在？或许，它的吸引力就在于与日常例行公事的生活完全不同？也许它是一种逃避？一种心爱的传统？很显然，我们看到的参加传统游戏和体育锻炼的大多数人都是老人。这是否意味着这些传统会消亡？还是随着年龄增长，这些传统会被年轻人继承发扬？本章将展示千禧年前后一个在中国城市的公共空间漫步者观察人们及其活动所能体验到的东西。

四川阆中有一座街巷狭窄、院落密布的老城，独立于新的城市发展区。中午和傍晚时分，走在安静的街道上，总会听到一种奇怪的声响，急促的哗啦哗啦声会在刹那淹没人们平静的交谈声。循着声音向一个院子中张望，你就会看到一群人在围着一张桌子打麻将，熟练而迅速地推倒和牌。人们用这种节奏清脆的麻将乐曲，为古城平添气氛。有人可能会以为，这种景象只是发生在阆中古城这样的传统院落环境中。其实不然，住在大都市高楼大厦中的人们，即使街区没有公共社会生活空间，也同样热衷这个游戏。如图07-3所示，居民们找到沿街窄长的一块地方，其空间刚好够安排一场麻将大赛。

棋牌游戏也随处可见，在公园，在小巷，在大街的人行道上。棋牌手们被兴致勃勃的观众团团围住，对比赛进程评头论足。

公园是"城市之肺"。高密度的住宅区需要户外休闲活动的开放空间。树木和其他绿色植物可以减少污染，提供新鲜空气。公园为集体活动提供了空间，也为那些寻求片刻宁静者提供了角落或隐蔽之所。在我持续走访（中国）城市的30年中，公园的作用没有多少改变。在向交通密集型道路、高层建筑、拥挤住宅区和繁忙街道生活发展的动态转变中，它们是永久的城市元素，也是一种缓解。

公园生活从清晨就开始了。你会遇到练太极拳和气功的人群，拿着五彩扇子跳舞的女士，在成年教练带领下做体操的孩子。在黎明和黄昏时分，你可能会听到从树丛中小广场传来的音乐声，人们翩翩起舞。留声机或 CD 机中的乐曲，有中国传统曲调，或西式老派快步和慢狐步舞曲。合唱团在排练，音乐爱好者骄傲地在公共场合演奏，或害羞地躲在树丛中练习。小学生和大学生们或坐在长椅上或走动着，大声朗读着课本，练习着英文。家人带着小孩玩耍。老人们锻炼身体或打牌。退休男子带着笼中鸣禽。这是一种美好的人类生活的嘈杂。幸运的是，这也是中国城市相当持久的文化特色。

图 .07-1
街边麻将
（西安，2013年）

Figure.07-1
Streetside
Mah-jong
Xi'an (2013)

图 .07-2
玩麻将的人们
（阆中，2014年）

Figure.07-2
Mah-jong
players
Langzhong
(2014)

图 .07-3
街边玩麻将
的人们
（西安，2013年）

Figure.07-3
Roadside
Mah-jong
players
Xi'an (2013)

图 .07-4
诊所外在休息
时间打牌的人
（柞水凤凰镇，
2007年）

Figure.07-4
Playing cards
during a break
outside the
local doctor's
office
Fenghuang
(2007)

图 .07-5
胡同里
打牌的人们
（北京，2012年）

Figure.07-5
Card players
in a *hutong*
Beijing (2012)

图 .07-6
古街下棋人
（柞水凤凰镇，
2007年）

Figure.07-6
Chess on the
old street
Fenghuang
(2007)

144

图 .07-9　　　　Figure.07-9
垂钓者　　　　　Fishing
（西安，2012年）　Xi'an (2012)

图 .07-7/8　　　Figure.07-7/8
早上七点　　　　Ballroom
跳交谊舞的人　　dancing at
（西安，1989年）　seven in the
　　　　　　　　morning
　　　　　　　　Xi'an (1989)

图 .07-10　　　　Figure.07-10
垂钓者　　　　　Anglers
（北京，1994年）　Beijing (1994)

图 .07-11
彩扇舞
（西安，2012年）

Figure.07-11
Dancing with
colorful fans
Xi'an (2012)

图 .07-14
购物后伸展腰腿
（北京，2013年）

Figure.07-14
After
shopping
outstretch
Beijing (2013)

图 .07-12/13
鸟儿和遛鸟人享
受新鲜空气
（北京，2012年）

Figure.07-12/13
Fresh air for
the birds–and
for the owners
Beijing (2012)

图 .07-15/16
日常健身
（北京，2014年）

Figure.07-15/16
Keep-fit
exercises
Beijing (2014)

146

图 .07-17
音乐排练
（北京，2014年）

Figure.07-17
Music
rehearsals
Beijing (2014)

图 .07-18
萨克斯管练习者
（西安，2011年）

Figure.07-18
Saxophonist
in a public
park
Xi'an (2011)

图 .07-19
公园乐器练习者
（泉州，1994年）

Figure.07-19
Exercise,
Quanzhou
(1994)

图 .07-20
湖上冰嬉
（北京，2008年）

Figure.07-20
Skating on the
lakes in the
city center
Beijing (2008)

Over the whole period of my work in China, it has been a pleasure to observe the life taking place in the public sphere, in courtyards and parks, on streets, and on sidewalks. For a foreigner, it's quite exotic and attractive to see these popular activities, which over time don't change much. They seem to be expressions of genuine culture and, unconcerned by tourist and commercial interests, they prevail. A friend of mine once remarked: It is vitalizing to observe, and it renews my belief in the human being.

As we observed China undergoing rapid and huge-scale changes, it was interesting to notice how some things seemed more permanent. Also changing, also developing, but changing at a different pace than the rest. In this chapter, I will show, with more illustrations than in the other chapters, some examples of these things that don't change so fast. What can this permanence tell us? Do that people need something rather permanent in their existence? Perhaps there are things that are more profound, some things that have a deeper appeal, and these things are harder to let go of. What then would we mean by "deeper appeal"? Appealing, perhaps, to many sides of human existence, more faceted than those things we deal with in day-to-day routines? Or maybe they are completely different from the day-to-day routines? Are they an escape, maybe? Or a beloved tradition? Evidently, the majority of the people taking part in the traditional games and physical activities are of the older generation. Does this mean that the traditions will die out? Or will they be picked up and evolved by the younger people as they grow older? This chapter will demonstrate what one can experience by sauntering in the public space, watching people and what they do.

Langzhong in Sichuan Province has, separate from the new, urban development, an Old City with narrow streets and courtyards. Walking in the quiet streets at lunchtime and in the late afternoon, you will hear strange sounds. Rapid clicks drown out the peaceful sound of human talk. Peeping into a courtyard from where the clicks seem to originate, you will probably see a group of persons around a table, playing mah-jong. Skilled and rapidly they set down their pieces. There are many groups of players like that, coloring the atmosphere in the Old City with this crisp rhythm, mah-jong music. One could believe that the phenomenon is taking place just in a traditional village environment like the Old City of Langzhong. But no. People resettled to urban, high-rise blocks in the bigger city, and also play the game, even if the block has no public space for social life. Have a look at the picture below where residents have found a stripe along the street outside the skyscrapers with just enough space for arranging a mah-jong tournament.

Also, chess and cards are played everywhere, in the parks, in small streets, and on the sidewalk of bigger streets. Players are surrounded by interested spectators, commenting on the drama of the match.

The parks are the lungs of the city. High-density housing areas need open space for outdoor, leisure activities. Trees and other greeneries damp down pollution and provide better air for human lungs. Parks make space for group activities, and they also make nooks or hidden spots for those who seek a moment of solitude. The role of the parks has been more or less unchanged during the 30 years of my continual visits to the cities. They have been permanent urban elements, a relief in the otherwise dynamic transformation towards traffic-intensive roads, high-rise buildings, congested dwelling areas, and hectic street life.

Park life begins early in the morning. You meet groups practicing Tai Chi Chuan and Qigong, ladies dancing with colorful fans, and children doing gymnastics led by a grown-up instructor. You may, at dawn and at dusk, hear music from a small square among the trees where elegant ballroom dancing is performed to music from a gramophone or CD player, traditional Chinese tunes, or Western-influenced, old-fashioned quick-step and slow-fox dance music. Choirs have their rehearsals, proudly some musicians play in public, or more shyly, hidden in the bushes. Schoolchildren and students practice oral English, reading textbooks out loud, seated on benches, or walking around. Families bring their small children for playing. Old people exercise or play cards. Retired men bring their cages with songbirds. It's a beautiful cacophony of human life. And, fortunately, it's a rather permanent cultural feature in Chinese cities.

08

50年代
的记忆
西安
2005

08

MEMORIES
OF THE
1950S
Xi'an
2005

1985年时，西安的街面很安静。驴子、马匹运送着钢筋和混凝预制板之类的建筑材料。路上有卡车、公共汽车，一些出租车，以及无数自行车。骑自行车既方便又安全。20年后，情况则已大不相同。新建的宽阔道路充斥着机动车，交通经常堵塞。骑车已变得极其危险。在西安，虽然有自行车道，但经常被停放的汽车占用，留给自行车的空间很小。任何骑车人都必须绕开停放的汽车，这是危险的活计，因为意味着骑车人要在汽车行列中进进出出。在西安的街道上，越来越少看到骑自行车的人。北京的长安街则不太一样，那里主干道很宽，每个方向都能容纳四五条车道，两边有自行车道，还有宽阔的人行道。这样的街道，在我们认知中更像是两端无限延伸的城市广场。而在北京的其他街道，和上面关于西安的描述大致相同。

为什么会出现这种交通模式的彻底改变？原因当然有很多：城市人口的快速增长，交通技术的发展，私家车数量的增加，城市的蔓延需要新的基础设施模式，诸如此类，不一而足。这里我想重点讨论的是，住房政策的变化如何影响交通模式。

20世纪80年代以前，职工住房由工作单位低价提供。住房改革后，工作单位的这一责任终止了。从80年代开始，家庭住房从工作单位拥有（产权）的福利品，变成由家庭自己支付、私人拥有的商品。此前，这些家庭住在工作场所附近，对机动交通的需求很小。现在，这些家庭出于不同原因——在急速扩张的城市中四处定居。结果在住房与工作单位间的距离问题上，形成（与此前）全然不同的本土化结构。反过来，又导致交通需求的大量增长。

我们一次硕士生课程的选址，在西安的西部城郊工业区。20世纪50年代建设时，工厂在选址和组织铁路公路交通方面拥有优先权。居住区建在工厂附近，人们可以步行或骑自行车来往于工作地和家庭之间。住宅区附近有必要的服务设施——幼儿园、学校、饭堂、公共浴池和商店——由此或多或少形成了对机动车交通需求较低的自治社区。当时这是一个非常合理明智的配置，当我们念及今天的污染和拥堵的城市交通时，这是一个非常值得记取的事实。

现在我们需要考虑的一个问题是，这些有人居住的历史文化遗产的两难困境。与后来建设的住宅区相比，其住房标准较低，缺少维护，给居民增加了负担。从好的方面来说，尽管有所减弱，社区原有的坚实社会网络随着时间的推移也在发展。解决住房标准低的一个简单办法，就是以新的现代住宅取代旧住宅区。由于这些新住宅在空间布局上有很大不同，其结果之一是形成了新的社群。在20世纪五六十年代，新的工业区建起时，就出现了这种情况。传统的城市院落生活被三五层高的苏式住宅区内新的社区生活所取代。当时的口号是"先生产后生活"，换句话说，其言下之意是适当降低住房标准。再后来，在80年代之后，又发生了一次类似的转变。这一次是商品住房和高层建筑占据主导地位，扮演了20世纪五六十年代的工业居住区的角色。

在我的家乡，我可以看到相似之处，一些开发商希望拆除旧的、衰败中的居住区，为现代建筑和道路开辟空间。而许多居民则对自己曾在恶劣居住条件下生活的老房子感情复杂。这样的住房可能是对贫困屈辱生活的一种提示，对于个体来说，这代表了他们人生经历中好和坏的部分。此外，这样的场所也有着更为宏观的角色——它们是我们集体记忆的一部分，是我们文化遗产的一部分。我们所有人，从普通市民到政策制定者，都喜欢记住好的东西，忘记坏的东西。当我们回忆自己的家园时，时间愈久远，我们往往会有更温暖的感触。时间愈近，我们越是倾向关注负面的东西，忽视或遗忘好的东西。

这些建筑代表着中国历史上一些重要的时期。考虑及此，就有一个非常有趣的问题：为什么我们不把20世纪五六十年代的城市住宅视为重要的历史？为什么人们对呵护、改善、更新它的兴趣如此之低？应如何尊重这一段对于中国居住历史（乃至中国历史）有着重要意义的时期？应如何改善这些建筑的居住条件，同时又能尊重其原有的价值与品质？如果我们想提升社会对修复这些建筑的兴趣，如果我们想让居民留在他们的单元房，找到最后这个问题的答案可能更为关键。利用创造性的思路和投资，改造翻新现有住宅，使其适应现代生活和现代人的期望，这会很有意思。找到解决之道并不容易，但探索总是让人兴奋。

图 .08-1
西安50年代
老住宅区
（西安，2005年）

Figure.08-1
Old residential
area in the
1950s
Xi'an (2005)

图 .08-2
北京长安街
（北京，1994年）

Figure.08-2
Chang'an
Avenue
Beijing (1994)

图 .08-3
1985年的
西安街面
（西安，1985年）

Figure.08-3
Street scene
Xi'an (1985)

图 .08-4
2005年的
西安街面
（西安，2005年）

Figure.08-4
Street scene
Xi'an (2005)

图 .08-5
50年代
住宅厨房
（西安，2005年）

Figure.08-5
Kitchen of a
1950s flat
Xi'an (2005)

图 .08-6
50年代
住宅外观
（西安，2005年）

Figure.08-6
Façade of
1950s housing
Xi'an (2005)

图 .08-7
邻里交往
（西安，2005年）

Figure.08-7
Space
for social
networking
Xi'an (2005)

图 .08-8
翻新的阳台空间
（西安，2005年）

Figure.08-8
Space for
refurbishing
Xi'an (2005)

In 1985, the traffic was calm. Horses and donkeys transport-
ed building elements, like steel beams and prefabricated,
concrete floor modules. There were lorries and buses, some
taxis, and not least, myriads of bicycles. Bicycling was comfortable and safe. 20 years later the picture was very different. Motorized vehicles filled the new and wide roads, often
creating traffic jams. Bicycling had become mortally dangerous. In Xi'an, there were bicycle lanes, but they were often occupied by parked cars, leaving little space for the
two-wheelers. Anybody still insisting to move around by
bike had to circumcise the parked cars, which meant going in and out of the motor-car files. It was a dangerous undertaking. Fewer and fewer bicyclists were to be seen in the
streets of Xi'an. It's different in Chang'an Avenue in Beijing,
which is wide enough to accommodate five to six car lanes
in each direction, bicycle lanes on both sides, and wide sidewalks. This street is more like a very long and continuous urban squares. Otherwise smaller streets in Beijing more and
less have the same conditions as mentioned above.

Why this total change in traffic patterns? There are many
reasons, of course: The rapid and dynamic growth of the urban population. Development of transport technology. The
increasing volume of private cars. Extended use of urban
territory needs new infrastructure models. And so on, and
so on. What I would like to focus on here is how changes in
housing politics affected the traffic pattern.

Before the 1980s, the workers were given their dwellings for
a low price by the work unit. This responsibility of the work
units came to an end during a housing reform. From the
1980s and onwards, the family's apartment changed from
being a welfare good, owned by the work unit, to a private-
ly owned marked commodity, paid for by the families them-

selves. Up until then, the families had lived adjacent to the place where they worked, with minimal need for motorized transport. Now the families – for different reasons – settled everywhere in the bulging city. The consequence, when it came to the distance between housing and work units, was a completely different localization structure. This, in turn, required extensive growth in transport needs.

The site for a master course we carried out, was in the western, industrial suburbs of Xi'an. At the time when they were established, in the 1950s, the industry had first pick in selecting a location and in organizing transport by train and roads. The housing areas were built near the factory sites, making it possible to move between job and home by foot, or by bicycle. The housing area had the necessary services close by – kindergarten, school, laundry, public bath, and shops – and thus more or less formed autonomous neighborhoods with low demands for motorized transport. At the time this was a very rational and sensible configuration, a fact well worth remembering when we think of today's pollution and congested urban traffic.

A topic we should consider is the dilemmas of this inhabited, historic cultural heritage. The housing standards were low compared to the housing areas built later, and the lack of maintenance added burdens to the residents. On the upside, although fading in the last years, solid, social networks have developed over time. An easy solution to the low housing standards was to get rid of old housing districts and replace them with modern homes. Since these new homes had very different spatial constellations, one consequence was new social groupings. This was the case in the 1950s and 1960s when new industrial districts were established. Life in traditional, urban courtyards was replaced by the new life in rational, "Soviet-style," three to five-story tall housing blocks. The slogan was "production first, livelihood second" – in other words, the implication was that the housing standard was intentionally low. And later, in the 1980s and onwards, a similar shift took place. Market housing and high-rise constructions dominated replacing housing blocks of the 1950s and 1960s.

I can see resemblances to my hometown, where some developers wanted to remove old decaying housing areas to create space for modern buildings and roads. Many residents expressed mixed feelings about their old homes, where they had lived under poor housing conditions. And while such homes may be a reminder of life under poor and humiliating conditions, and represent both good and bad sections of their personal biography, such districts play a more comprehensive role. They are part of our collective memory, a part of our cultural heritage. We all, from plain citizens to policy-makers, like to remember what was good and repress what was bad. As we remember our homes, the longer ago it was, the warmer feelings we often have. The closer in time, the more we tend to focus on the negative and suppress or forget what was good.

These areas represent important periods of Chinese history. With that in mind, very interesting questions occur: Why don't we consider the urban housing of the 1950s and 1960s as historically important? Why is there such low interest in taking care of it, improving and modernizing it? What could be done to pay respect to this important period of Chinese housing history, nay Chinese history in general? What could be done to improve the housing conditions in these buildings and at the same time respect the original qualities and values? Finding answers to the last question is probably the key if we seek to increase the government's interest in restoring these buildings and if we want the residents to stay in their apartments. It would be interesting, to use creative ideas and investments, to modify and refurbish the existing housing volumes, and to adapt them to modern life and modern-day expectations. It will not be easy to find solutions, but it is very exciting to search for them.

09

全球化
与研究
2011—
2013

09

GLOBALI-
ZATION
AND
STUDIES
2011-
2013

关于真实性

也许，作为一个起点，我们可以提出一个非常笼统、非常哲学的问题：是否存在普遍的真实性？大多跨文化的对话都基于这样一种信念：不同国家或族群尽管利益取向不同，文化、宗教信仰、物质标准也不尽相同，但一定有一些共同的东西。毕竟，没有这点，对话无从谈起。人类学家接触研究不同社会中的人群，并依靠他们的专业方法建立起对各种文化的深刻洞察。一些人类学家试图在众多差异的背后找到一些潜在的共同结构。例如，在人类体质结构中是否有共同的因素引导着人类的行为？即人类学家克洛德·列维-施特劳斯（Claude Levi-Strauss）的理论——文化模式由人类思维中潜在的心理结构演变而来。也许这又根植于人类大脑中的生物结构，谁知道呢？无论如何，如我们在教授中国学生并与之交流时所体会到的，跨文化的对话的确是很有趣的过程。

历史上不乏直接和间接文化殖民的例子。强权、强势的声音压倒弱势一方，把自己的一套价值观视作普遍真理。二战后，建立了一系列全球性组织和国际公约，力图在国际交流互动中找到共同价值，从而推动和平。而在过去几十年里，公共舆论则批判地指出，这些安排是否真的建立在普世观念基础之上？还是强国无意识甚至有意识地以隐蔽的议程所打造安排？或者，一种不那么偏激的说法：也许一些最初被确信为普世的价值观，事后发现并非如此？我想到的例子是联合国大会在第二次世界大战后不久通过的《世界人权宣言》。宣言的本意是要促进世界上所有人类和平与平等地生活，但在实际政治运作中，它却成为一个如何解释的问题。类似地，我还想到了联合国教科文组织和国际古迹遗址理事会关于遗产保护的理论性文件。例如，1964年的《威尼斯宪章》表达了一些绝对和普遍性规则，后来又经修改，以涵盖最初宪章起草者未考虑到的其他情况。有些人认为，编写该宪章的委员会由意大利专业人士主导，他们基于意大利丰富的砖石建筑遗产经验对宪章的行文影响甚大。1994年的《奈良真实性文件》通过引入与地方性条件相关的规则，使得宪章文本用词变得缓和。虽然最初的协议是为了达到理想和公正的目的，但不可避免的是，后来围绕文本调整进行的讨论成为政治性的。行文至此，就让我把这些跨文化交流的反思和对国外学生的教学联系起来。

教授中国学生

2011—2013年，西安建筑科技大学聘请作为建筑师 / 遗产管理者的玛丽娅（Marie Louise Anker）和我为中国建筑专业学生开设了两门课程，一门是关于居住和住宅理论，另一门是关于保护、文脉和环境。我们的主要想法是将这两门课程联系在一起，将其纳入在地或本土建筑学的范畴。中国同事邀请我们从欧洲 / 挪威视角提出问题，因而课程架构也相应地基于以下主题。

以建立一种批判性方法为目标，我们首先讨论了一些适用于所有具体议题的总体性问题。例如，在讨论遗产保护时，一个关键问题是为什么要保护？这个问题的讨论，涉及处理记忆、历史和遗产之间的联系，以及其作为住居发展和身份形成的要素等几方面。我们还批判性地讨论了关于保护的概念和宪章公约。在类似的架构下，居住理论课包括了全球视野下的居住、生活方式和住宅标准等主题。方法论方面的问题则单独予以讨论——如一般研究方法，特别是人类学方法，以及围绕使用者参与的各种问题。最后，对如何在特定机理环境下进行规划设计进行了回顾，例如，如何在旧环境中建造新建筑（对比抑或适应）；如何与物质环境相关联，无论是人为抑或自然环境（地形、气候等）；如何通过节约和有效利用资源尊重环境；如何与社会文化语境相联系。

研讨讲座以对谈和讨论为基础，课程考试是论文写作。这提供了一个宝贵的视角，了解年轻的中国学子对世界的总体看法，以及他们个人的居住体验。这也很好地反映出他们对自己文化遗产的态度，特别是建成环境中的遗产。工作语言是英语，这当然会导致一些沟通上的问题，因为没有人的母语是英语，并且英语水平参差不齐。不过，这也是一个启蒙的过程，至少对教师而言如此。在课程期间和课程结束后，我的脑海总是会浮现出一些批判性的想法：我们是否把自己的理想、自己的生活方式和思考强加给学生？我们是否像以前的传教士或今天的国际媒体那样在传播福音？的确，我们之间的对话永远不可能发生在一个理想化的公平竞争的环境中。教师总是有优先权，这不一定是智慧和经验的结果，而是因为教师的角色带有内在权威观念。这种不平衡很难消除。

除了在中国开展的教学，在挪威科技大学建筑学院，我们也很乐于接收中国学生进行博士学习。在某种程度上，他们与导师处于平等的位置——他们受过良好的教育，对感兴趣的领域有深入研究。而同时他们与大学签有工作合同，更多受新学习环境的学术文化左右。当中国博士生与挪威教授相遇，文化背景的差异也表现为他们在学术传统方面的差异。

自20世纪90年代起，一些来自西安的年轻同事在挪威科技大学攻读了博士学位，将他们的研究与我们大学的教学和实际项目联系起来。董卫研究了西安鼓楼回民区的改造。王韬对中国20世纪50年代以来的住房政策和住房供应的发展进行了考察反思。许东明比较了中国和挪威对考古遗址博物馆的做法，分析了基本价值取向的差异。王宇研究了地震灾害后传统村落的重建问题。王燚通过研究公共空间的使用变化，关注日益增长的旅游产业对历史村落凤凰古镇的影响。他们共同为文化遗产保护与发展领域做出了坚实有价值的贡献。*10
他们的一些研究是对本书所述一些主题的直接延伸和深入考察。

传统与创造

从一开始，中欧学校系统的教学方法就是不同的。以教孩子读书写字为例：中国的小学生要记诵无数汉字，通过反复重复老师所说所写，一个个字用心汇集；而挪威的孩子只需要学习29个字母，怎样写和发音、怎样组合，就可以自由阅读、遣词造句。我相信这种差异对于思维的进一步发展是非常基础的。它在头脑中形成模式，当这些个体面临解决新问题的挑战时，这种模式就会显现。中国学生，我只认识有限几个，我并未对此做过深入观察。当进入新的领域，榜样化的实例不在眼前，需要其创造力和想象力的时候，他们需要迈过更高的门槛。

＊10　董卫. 转型中的民族住居——资源管理与场所认同框架下的中国穆斯林居住建筑 [D]. 挪威理工学院1995年；王韬. 中国城市住房供给制度改革的社会视角：北京、西安和深圳的三则案例 [D]. 挪威科技大学2004年；王宇. 如何重建受灾遗产聚落：桃坪村震后重建研究 [D]. 挪威科技大学2015年；许东明. 遗址博物馆多维视野考察 [D]. 挪威科技大学2018年；王燚. 旅游对中国凤凰古镇的影响，以及建筑与公共空间的变化 [D]. 挪威科技大学2019年。

一个相关的例子是我们如何理解创造。在多年前的一次会议上，一位演讲者介绍了一个理论，尽管有些卡通化，但我觉得非常有趣。他说在中国，创造的意思是在已知和已接受的基础上，小心翼翼地添加一些东西。在西方国家，创造则被认为是创造全新的东西。演讲者认为，前者是由儒家基于礼教、纲常、孝顺的思想所致；而后者则与基督教认为上帝从无到有创造一切的信仰有关。不管对不对，至少对于我这个不乏偏见的人来说，这个理论的确指向了文化差异背后更为深层的结构。

另一个有趣的问题，就是理解和处理概念的差异。各个学术领域都有许多例外和不同的理解与做法，但为了便于论证，我还是以偏概全：在西方，尽可能精准地定义一个概念是一种美德。而中国文化中，对文本的表述模棱两可，为各种理解和阐释留有余地，是一种优点。在博士研究工作中，我们能否尊重这两种倾向？我们又应如何包容差异？我的一个博士生说过，必须要"以诗译诗"。作为博士生，在异国他乡很难去兼收这些差异，因为在异国的学术环境中去学习，有其他的原则、学术规范和实践做法必须遵循。

走笔至此，我暗自怀疑这一切在近些年已然发生变化。也许出国留学学生人数的扩大，以及研究项目国际化程度的不断提高，有助于减少这种差异。无论如何，既然情况可能已然发生变化，就把这段文字作为我与中国师生最初几年接触的印象记录吧。

图 .09-1
午间放学的校园
（西安，2012年）

Figure.09-1
Campus after
school at
lunchbreak
Xi'an (2012)

图 .09-2
勤奋而有激情
的学生
（西安，2012年）

Figure.09-2
Enthusiastic
and hard-
working
students
Xi'an (2012)

图.09-3
这是一个漫长艰苦学习历程的开始：也许建构理解和知识的基础，对中国孩子来说，是从学习书写和阅读复杂多样的中国字词开始的；这与西方孩子在学会二十几个拉丁字母即能开始阅读和书写的情形非常不同
（柞水凤凰镇，2009年）

Figure.09-3
The start of a long and laborious process; maybe structuring the basis for understanding and knowledge is, for the Chinese children, formed from the very beginning of learning to write and read the diversity of Chinese letters; very different from our children who can read and write when they have learned the few letters of the Latin alphabet Fenghuang (2009)

On verities

Perhaps, as a starting point, we could pose a very general and philosophical question: Are there universal verities? Most international dialogues are based on a belief that despite different interests, different cultures, religious beliefs, and material standards, there must be something common. After all, a dialogue would be impossible without it. Anthropologists meet and study populations in different societies, and leaning on the methods of their profession they build a deep insight into a variety of cultures. Some anthropologists try to find something underlying, some common structure behind the many differences. Are there, for example, common elements in the physiology of human beings that steer human behavior? This is what anthropologist Claude Levi Strauss theorized: That cultural patterns evolve from underlying mental structures in human thinking. Maybe this again is rooted in structures in the human brain. Who knows? In any case, dialogue across cultures is interesting, as we experienced when we taught and talked with Chinese students.

History is full of examples of direct and indirect cultural colonization. Strong powers and strong voices overrule the weaker ones, posing their own set of values as universal truths. After World War II, a set of global organizations and international agreements were established, seeking to find common values in international interaction and thus encourage peace. During the last decades, public opinion has asked, critically, whether these arrangements really are based on universal ideas-or are strong nations forging them, unconsciously or even consciously with hidden agendas. Or, less inquisitorial, the question has been, if maybe some values that initially were assuredly assumed to be universal, in hindsight turn out not so much so. I have in mind, for example, the Universal Declaration of Human Rights adopted by the United Nations General Assembly shortly after World War II. The intention was that the Declaration should promote peace and equal living conditions for all people in the world, but in practical politics, it becomes a problem of interpretation. I also have in mind doctrinal texts on conservation from UNESCO and ICOMOS. The Venice charter of 1964, for example, expressed some absolute and universal rules, which were later modified to cover other situations originally not considered by the authors of the original charter. (Some say the Committee who wrote it was dominated by the Italian professionals whose perspective of Italy's rich stone architecture strongly influenced the wording of the charter). The Nara documents of 1994 softened the charter texts by introducing rules relative to local conditions. While the original agreements are intended to be ideal and impartial, it is unavoidable that discussions around later adjustments in the texts turn political, not ideal and impartial as the intention might be. Let this so far be a reflection linked to the teaching of foreign students.

Teaching Chinese students

During a period of three years (2011-2013), XAUAT engaged architect/cultural heritage manager Marie Louise Anker and me to give courses for Chinese architect students, one in the theory of housing and dwelling, the other on conservation, context, and setting. The main idea was to link both courses under the umbrella of sited, or localized architecture. We were asked and challenged by our Chinese colleagues to present problems from a European/Norwegian perspective, and the program was built up accordingly with topics as follows.

Aiming for a critical approach we started by discussing some overall problems applicable to all the specific topics. When discussing conservation, for example, one crucial question was *why conservation*? This question was discussed by handling aspects like the link between memory and history and heritage as elements in housing development and identity formation. Critically notions and charters on conservation were discussed. The class in housing theory included, against an analogous backcloth, themes like housing in a global perspective, and housing, lifestyle, and standard. Methodological aspects were addressed separately – as methods in general, anthropological methods specifically, and challenges around user participation. Finally, how to practice contextual planning and design was reviewed, like how to build new in old settings (contrast or adaptation?); how to relate to the physical context, be it man-made or natural (topography, climate, etc.); how to respect the environment by frugal and efficient use of resources; how to relate to the socio-cultural context.

The seminars were based on dialogues and discussions, and the exams were on essay-writing. This gave a valuable insight into how young, Chinese students conceived the world in general, as well as their personal housing experiences. It also gave a good picture of their attitude towards their cultural heritage, especially in the built environment. The working language was English, which of course caused some communication problems since no one's mother language was English, and the English skills varied a lot. Nevertheless, it was a process of enlightenment, not the least for the teachers. During the courses and afterward, some critical thoughts cropped up: Were we imposing our own ideals, our own lifestyle, and thinking onto the students?

Were we evangelizing like the old missionaries used to do, and like the international media now does? Sure, we had our dialogues, but this could never be on a level playfield. A teacher will always have precedence, not necessarily a result of intelligence and experience, but because the role of - the teacher comes with a built - in idea of authority. It is hard to eliminate this imbalance.

In our faculty, we have had the pleasure of inviting Chinese students to carry out their doctorate studies. They are on somehow equal terms with their supervisors - they are well educated and delve deeply into a field of interest. But they are simultaneously through their contract with the university, more at the mercy of their new study environment's academic culture. When Chinese candidates meet Norwegian professors, differences in the cultural background also reveal themselves as differences in academic traditions.

Some young colleagues from Xi'an have done doctorate studies at our university, NTNU, linking their research to our teaching and real projects: Dong Wei studied the transformation of DTMD in Xi'an. Dongming has compared Chinese and Norwegian approaches to site museums, analyzing differences in underlying values. Wang Tao reflected on the housing policy and the development in housing supply since the 1950s. Wang Yu has studied the re-establishment of traditional villages after earthquake catastrophes. Wang Yi has focused on the effect of growing tourism in Fenghuang historic village, by studying the changes of use in the public realm. Altogether they have given substantial and valuable contributions to the field of cultural heritage protection and development. * 4 Some of their studies were direct prolongations and deeper investigations of themes touched upon by other studies described in this book.

Tradition and creativity

From the very beginning, the pedagogics of the school systems are different. Take for example how children are taught to read and write. The Chinese pupil has to learn innumerable characters by heart, meticulously collecting them, one by one, by repeating again and again what the teacher says and writes. The Norwegian child only has to learn 29 letters, how to write and pronounce them and how to put them together, and then they are free to read and formulate texts. I believe this difference is very basic for the further development of thinking. It creates patterns in the minds, which become visible when the individuals are being challenged to solve new problems. Chinese students – and I'll hasten to point out I have known only a few and thus I have not investigated this deeply – have

---

* 4   Dong Wei. "An Ethnic Housing in Transition. Chinese Muslim Housing Architecture in the Framework of Resource Management and Identity of Place." Ph.D. Diss., NTH, 1995; Wang Tao. "A Social Perspective on the Reformed Urban Housing Provision System in China: Three Cases in Beijing, Xi'an and Shenzhen" Ph.D. Diss., NTNU, 2004; Wang Yu. "How Do We Rebuild a Disaster Damaged Heritage Settlement: A study of the Post-Earthquake Reconstruction of the Village of Taoping. A Traditional Qiang Settlement in Sichuan China." Ph.D. Diss., NTNU, 2015; Xu Dongming. "A Multi-Perspective Observation of Site Museums. Case study of Archaeological Site Museums in China, with Norwegian Example as Reference." Ph.D. Diss., NTNU, 2018; Wang Yi. "The Impact of Tourism in the Historic Town of Fenghuang, China and the Changes in Buildings and Inter-Building Public Spaces." Ph.D. Diss., NTNU, 2019.

strong abilities to collect empirical material and at the same time have a higher threshold to pass when new land is entered when model examples are not at hand and creativity and fantasy is challenged.

A related phenomenon is how we understand creativity. A speaker at a conference years ago introduced a theory, which, despite the fact that it is somewhat caricatural, I found very interesting: In China, creativity means to marginally and carefully add to what is already known and accepted. In Western countries creativity is conceived as creating something completely new. The speaker argued that the former is caused by Confucian thinking – based on rituals and conventional, filial obedience; while the latter is linked to the Christian belief that God created everything out of nothing – True or not, at least the theory points to something I, perhaps in my prejudiced mind, have felt as deep-structured phenomena.

Another interesting issue that showed up was the differences in understanding and handling notions. There are many exceptions and diverse understandings and practices in various academic fields, but for this argument's sake, I will generalize: In Western countries, it is a virtue to define a notion as precisely and exactly as possible. In Chinese culture, it is a virtue to formulate texts ambiguously, giving many openings for various understandings and interpretations. Could we respect both tendencies in the doctorate work? How could we embrace the differences? One of my doctorate candidates said that one had to: "translate a poem with a poem." As a Ph.D. student it is hard to navigate these differences in a foreign country, where there are other ideals, and other rules and practices in academic work.

176

As I write this, I get a sneaking suspicion that all this has changed in recent years. Maybe the extended volume of students going abroad for their studies and the increased internationalization of research projects contribute to diminishing such differences. At any rate, if things have changed, let this be a record of my own impressions during the first years of contact with Chinese teachers and students.

10

参与的困境
西安
2008/
2011—
2014

10

DILEMMAS OF INVOLVE-MENT
Xi'an
2008/
2011-
2014

印度非暴力自由斗士、圣雄甘地曾说："我不相信最大多数人代表着最大化的利益理论。它赤裸裸地意味着，为了实现所谓51%的利益，可以或者说应该牺牲49%的利益。这是一种无情的理论，对人类造成了伤害。唯一真正的、有尊严的人类理论是对所有人的最大利益。"*11

圣雄甘地理想的不妥协态度让人感动。相信每个人都想按照甘地的原则生活，但很多人可能同样会觉得这理想有些幼稚，当然也很难实践。只要有大量的人参与到一个过程中，就会有矛盾的需求和梦想。那么，我们如何做出最终的决定？我们如何了解参与者，即所谓利益相关者的各种意见和优先事项？在这个过程的最后，总会有某种决策者。他们对人们的需求有什么样的了解，他们如何获得这些知识以及如何与之相关联？本章将在历史街区的保护与发展领域对这些问题进行思考，首先从国际、国内章程审视，然后就挪威科技大学参与的两个案例探讨这些问题。其中一个是西安的唐大明宫遗址考古公园国际竞赛；另一个是同样在西安的汉长安城未央宫遗址研究项目。

在中国，对文化遗产关注的范围，无论对于什么是文化遗产，还是如何处理文化遗产，大体上是随着国际学界发展的。其国内实践与国际实践的主要区别，似乎在于普通人对遗产地的保护和发展过程的参与。当国际规范遇到国内立法，这种差异就变得至关重要，因为中国的项目需要根据国际规范进行评估，例如在申请将遗产地纳入联合国教科文组织的世界遗产名录时。而在名为《中国文物古迹保护准则》的文件中（简称《中国准则》，初版是在2000年，2015年修订）*12，其第八条关于参与的规定是："文物古迹的保护是一项社会事业，需要全社会的共同参与。全社会应当共享文物古迹保护的成果。"

从修订后的《中国准则》可看出，国家法规是跟着国际上的修订在发展的。那么，亟待解决的问题是如何阐释、遵守和践行这些条例。

---

*11　引自印度艾哈迈达巴德圣雄甘地纪念馆铭文。

*12　该文件由中国国家文物局与美国盖蒂保护所、澳大利亚遗产委员会合作编制，是一份协作完成、尊重并反映中国传统和方法的文化遗产保护国家纲要。

有人居住的历史街区的保护与发展总是充满妥协。国际国内法则的推荐意见描述了理想的条件和最佳做法。与其他国家一样，理论与实践之间存在着差距。有些调整可能是必要的，有些则没有必要。在下文中，我将详细阐释上述相关问题。

美与真的
两难

文化遗产保护中的一个难题是在审美方面。在这方面，主要涉及保护的基本意图。保护的主要目的是美化保护对象，还是让修缮显示出时光流转和岁月痕迹？哪个更为重要？这一点，保护工作者和建筑师们已讨论多年。无论是《威尼斯宪章》*13还是相抵牾的理论文本，如修订后的《中国准则》，都建议文物新添加部分应该是可识别的，而也有很多人认为这个原则有悖于文物的审美和谐。

我们于保护对象总有一个理想化的图像，而这个图像是由我们对其创建时状态的观念所塑造的。这种对客体原初状态的想法当然可能是基于有缺陷、不加批判的信息。这种两难还会出现在当我们采用成见去判断什么美、什么不美时。这时，求美往往可能意味着将所有我们认为令人不安、不规则的东西抹平。而出于我们对过去的理解，则似乎有必要修正我们的审美理想，把历史变迁的踪迹也包含其中，就像我们如何在一张老人脸上的皱纹中发现美一样。

这些复杂的问题展示了关于保护的两种主要不同态度：一种是清除与遗址创建初期有直接指向关系特征之外的任何内容；另一种是把肇始时代之后发生的一切都视为遗址历史的重要组成部分。后一种原则的结果，是显示具体的时间段落，即把遗址最初创建的时间作为历史事件连续性的一部分。换而言之，它表明该遗址的历史始于其最初创建时间之前，并通过时间推移至今。历史萦绕于斯，所有时期都留下印迹。

---

*13 《威尼斯宪章》是1964年由一批保护专家在威尼斯制定的一整套保护原则，为历史建筑的保护和修复工作提供了一个国际框架。在1965年被国际古迹遗址理事会正式采纳。

对原则的选择会产生实质性的后果。在大明宫和汉长安遗址案例中（见下文），都遵循了前一种原则，结果是当地村民被迁出，重新安置在其他地方。

体制的
两难

社会科学中，我们把治理分为两种：官僚体制，即根据法律规定人人享有平等的权利；以及侍从体制（通常称为"雇主—侍从"体制），即雇主获得侍从者的团结和支持，而侍从者通过交换获得利益。法治即官僚体制，而侍从关系则更灵活。侍从关系意味着权力被滥用的可能，但也意味着更灵活适应具体现实情况的可能。

中国的"关系"现象可以说是侍从关系的特殊版本，但并非一对一的关系，而是由几个人组成的关系网。按我的理解，关系网内有相互的义务。它是灵活的，但有不成文的协议和共识来规范什么该做，什么不该做。我对"关系"唯一的一次直接体验，是我们第一次来中国的时候，非常正面。我们在北京认识的一位大学同事，当他听说我们打算在中国国内旅游时，给了我一份他大学同学的名单，上面有他们的电话号码和生活、工作地信息。他告诉我们，必要时就和他们联系，并附上他的问候，如果有需要就请他们帮忙。我们确实与名单上的一些人取得了联系，此外，我们还在西安找到一位联系人，这成为我们今后30年交往的起点。

在规划和实施规划时，一个典型的两难情况是数量问题——在中国也如此。当受干预影响的家庭数量众多时，与每个家庭的直接沟通会占用很多资源。官僚体制的管理模式，通常是中国的首选模式，就像在其他国家一样，通常会减少矛盾和耗费时间的辩论。如果考虑每一家的需求和愿望，很容易造成不同家庭间的争风吃醋，或者——当改变发生时——让这些人家感觉受到不公正和不平等的对待（例如在征地和重新安置的情况下）*14。而激起这种情绪显然有违建立一个和谐邻里关系的目标。另一方面，考虑个体的情况，规

---

*14　摘自与一位汉长安城遗址保护规划编制者的谈话。

划前提可能会发生很大的变化，其他解决问题的方法可能会显现出来。举例来说，为了有效组织一项市区重建计划，明智的做法是先调查哪些人愿意留下来，哪些人愿意搬走。这个策略可以节省大量时间和其他资源。*15

让我们带着这些思考，看看大明宫和汉长安遗址案例的情况。

2007年

大明宫遗址

公园竞赛

唐长安的宫殿区位于西安老火车站以北，占地3.6平方公里，有农田、小型产业、居住区和宫殿建筑遗址等重要历史遗迹。作为1961年以来首批全国重点文物保护单位，作为中国帝制社会鼎盛时期唐文明的最具代表性所在，采取措施建设国家考古遗址公园的时机已然成熟。为了实现这一目标，第一阶段是集思广益，为该片区的未来规划提供指导。在2007年10月，西安市文物局和西安曲江大明宫遗址区保护改造办公室在上述框架下共同组织了一次国际竞赛。

经过首轮初选，八家团队的方案被选中参加最后的复审。其中一支团队来自挪威特隆赫姆，由挪威科技大学、BARK 建筑事务所和AGRAFF 景观与建筑事务所的建筑师和规划师组成。清华大学的一位中国同事有段时间也加入该团队工作。这样的组合让这个团队有了 "TEAM 3+" 这名字。

我们项目提案的基本优先事项是降低环境压力（对交通系统、绿地、建筑等）；精心管理考古构成内容（最小化发掘、筏板式基础轻质结构、保护现存考古地层）；维护现有住民居住环境；避免对过往未知结构的猜想性 "复制"。

---

*15 为使侍从制成为一种有建设性的选项，有必要对相关人的权力做更为密切的监督。

我们总体规划的主要内容参考了唐代宫殿的原有分区。南部是唐皇宫中的公共区域。项目提出在那里建一个城市发展区，保留和发展延续（现存）建筑体量中最有价值的部分，并对新增建筑进行规划控高。中部区域在唐代用于正式朝政管理——我们的项目提案建议它应该是一个知识区，包括博物馆、教育设施以及考古发掘材料的展览。北部为皇帝非正式朝务活动和家庭生活的区域——我们建议将其作为一个公园区。*16

我们选择重点关注南部区域20世纪形成的建筑和街巷。这里的建筑有的很破败，有的很完好，大多介于两者之间。这些建筑有三个方面的价值：它们是世代居住于此者的家园——其中一些来自河南，仍操着原来的乡音俚语；它们代表着中国历史的一段重要时期；最后，与唐代宫殿的纪念性规模形成对比，它们可以用来说明日常生活尺度结构和统治阶级建筑巨大规模尺度之间的关系。我们（由此）建议对现存建筑进行详细研究和评估。具体方法是与居住在该地段特别是南部区域的居民对话，找出哪些值得修复，哪些可以拆除——而不是将整个住区全部拆除。

最后所选取的策略则与此迥异：这个项目实施时，南部区域的所有建筑被拆除，以为考古公园腾出空间。*17

---

*16　译者注：其由南向北，分别对应以大明宫遗址三大殿所构成的"大朝""中朝""内朝"三部分空间。"大朝"以含元殿为主体，面朝丹凤门广场，国家盛大庆典多在此举行；"中朝"以宣政殿为主体，朝廷各重要机构如中书省、御史台等均设其左右，为皇帝处理日常朝政及百官办事的行政中心；"内朝"以紧连后宫的便殿紫宸殿为主体，官员在此殿朝见亦称"入阁"。北部为生活建筑区，以太液池为界分为东、西两大活动区。东部蓬莱阁、浴堂殿、绫绮殿等，为皇帝与后妃的活动区；西部以麟德殿、金銮殿和翰林院等为主，是皇帝在内廷引对臣僚，举行宴会和观乐赏戏之处。参见：史念海. 西安历史地图集 [M]. 西安：西安地图出版社，1996年，第88页。

*17　译者注：该区域20世纪棚户区大部分被拆除，仅在大福殿与麟德殿遗址间保留了18座院落，作为大明宫遗址公园的一部分予以改造利用。参见本书附录"刘克成、肖莉访谈：一位挪威建筑学者的人生启示课"。

汉长安遗址
研究项目
(2011—2014年)

2014年6月，丝绸之路的首段，即"长安—天山走廊"，被联合国教科文组织列入《世界遗产名录》(WHL)。西安作为东端的终点站，以未央宫遗址为代表，如图10-8所示，是整个汉长安城宫殿遗址区的几处内容之一。

汉长安城遗址位于西安市中心西北侧二环路外，总面积36.5平方公里，是一处受到城市扩张发展巨大压力的城区。这里有54个村庄居住着6万~7万人，其中许多是在宫城区域内拥有土地的农民。该区域对耕种有规定，限制深挖，以免破坏众多地下文化遗迹。这种规定对农民家庭的经济状况造成影响。自1959年始，该遗址进行了考古发掘，揭示了西汉时代生活的重要发现。但大部分遗址位于地表之下，仍未揭露挖出。与唐代一样，汉代也被认为是中国历史上的一个高峰期，故在地方和国家层面上，该遗址都是一个极具价值的标志。不断扩大的旅游业也对这片遗址兴趣浓厚。

这种情况复杂且难以管理。官方对此讨论经年。而现在，一个新的视角引入，为讨论增加了另外一个层次。如果要进入世界遗产名录，则很难绕开联合国教科文组织国际视野下的各种规范和公约。丝绸之路项目*18是跨国项目，这使得相关国家的程序、时间安排和流程变得更复杂。在严格的期限内，必须做出决定，以提交合适的申请材料。

在挪威驻华使馆赞助下，来自挪威科技大学和西安建筑科技大学的研究人员组成联合团队，在2011年至2014年的三年时间里，开展了一项名为"汉长安历史遗迹的保护与发展：当地利益相关者参与"的项目。团队采用以下几种方法研究这一案例：定期跟踪媒体报道（报纸、电视、广播、互联网）；采访利益相关方人员（村民、当地商户、地方领导、规划师、考古学者、历史学者等）；在现场调研

---

*18 一般被称为"序列性世界遗产"(Serial World Heritage)。

中进行个人观察；与项目的核心决策者、专业人员和规划师组成的参考组成员举行研讨会。

利益相关方的参与方式有很多。以下是一个例子。在汉长安城研究项目开始的前一年，挪威科技大学与西安建筑科技大学的教师合作，安排了一个硕士生教学课程，调查该地段的潜力，采用了几种参与方式。除提出设计方案外，学生们还组织了与村民接触交流的活动。例如，他们与当地一位村民一起安排了一次电影放映会，这位村民在过去经常在晚上放电影。他还保留着他的旧电影放映机和一堆老电影胶片。一面墙刷成白色作为屏幕。在漆黑的夜晚，电影放映吸引了大部分村民。另一支团队则从村里的学校请来一个班的学童，组织了一次寻宝活动。学生们要找到藏在汉代重要建筑遗址上的文物。他们还绘图作画，在室外平台上展出。

大刘寨村一位11岁的女生写道：

"我喜欢在田野奔跑，和朋友们一起玩耍。田野里有很多土丘，我们平时喜欢在上面爬来爬去。突然有一天，我们村西边的一个土丘被用木板盖住了。妈妈告诉我说，村子周围的土丘都是汉长安城的一部分。虽然现在已经是废墟了，但它们仍然是宝贝。这太神奇了！我们几个好朋友一起去看了被保护起来的土丘。它看上去很美，我们躺在木板上，感觉很棒。我希望这里的遗址都能得到保护。而我们的村子也会更美丽！"

为什么我们对当地利益相关方的参与感兴趣？联合国教科文组织和国际古迹遗址理事会的国际理论文件参照人权公约强调了这一方面的重要性。举例来说，申请列入《世界遗产名录》，必须记录当地人如何参与并获知情况，以及在今后的发展和管理中如何保持这种参与。联合国教科文组织在《实施〈世界遗产公约〉操作指南》中也强调了这点。

我们使用相关国际理论文本作为标准，观察和评估当地不同利益相关者如何参与，以及如何获知情况（什么时候，由谁告知）。我们选取了未央区的两个村子——大刘村和高庙村作为研究对象。项目期间，每两个月会与村民进行一次定期访谈。每到一处，我们都受到热情接待。当危及私人财产和未来生活时，这样的情况在世界任何地方都是敏感的。而由于情况敏感，尤其是第一年，我们对可能引入的话题做了一些限制。再后来我们感觉可以自由提及任何问题。有些受访者表达自己的看法非常谨慎，但我们与之交谈的大多数人，都喜欢谈论他们的日常生活，他们如何看待这种情况，以及他们对未来有哪些期望。

2012年12月至2013年2月，约有15000名村民被搬迁，9个村庄被拆除。2012年秋季，就在搬迁前，居民们才得到可靠的情况通告。在此之前，有关部门向村民提供的信息很少。他们所知道的都是基于口头传言或网络传言。他们无法知道自己村子的命运，也就很难规划自己的未来。

事实上，只有官员和专家参与了这个项目。当地的村民、商户，除了在他们不得不离开自己的家园、移走先人坟墓时参与了实际事务外，没有以其他任何方式参与。对于搬迁到（西安）三环外的高楼新区，各代人的感受不尽相同。一般来讲，老人们更愿意待在自己的老街，在那里可以和街坊邻居们聊天。年轻人则期待着改变和新的可能。换言之，年轻人对场所的依恋是不同的，由于对自己未来的不确定，也许他们对所在地的归属感在漫长的过程中逐渐消失了。

在本章，我试图说明，由于选择了这样的策略：在大明宫遗址中清除唐代以外的所有历史印迹，以及在汉长安城（未央宫）遗址中清除汉代之外的历史印迹，其结果是当地利益相关者的参与度的最小化。

图 .10-1
村民公共活动
（西安三原,
2012年）

Figure.10-1
A villagers'
public event in
Sanyuan
Xi'an (2012)

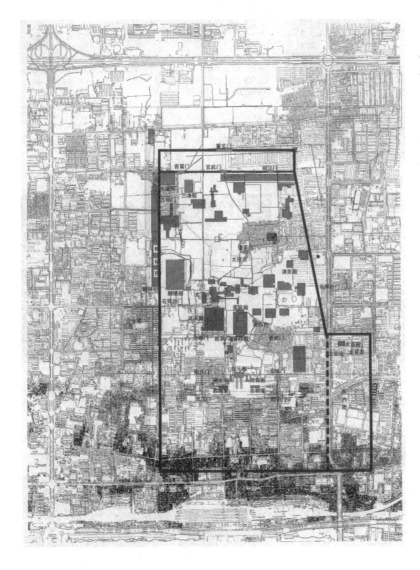

图 .10-2
唐大明宫遗址区
考古遗迹分布图

Figure.10-2
Cultural relics
on the site

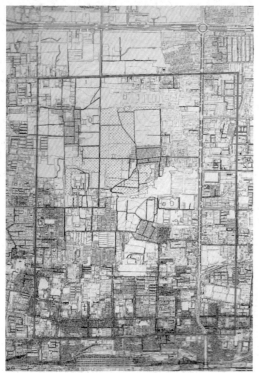

图 .10-3
唐大明宫遗址区
范围

（图源：西安大明宫国家遗址
公园概念设计国际竞赛公告附
件，2007年）

Figure.10-3
Boundaries of
the Daming
Palace site

(Photo credit: Competition
material – international
conception design
competition on Xi'an
Daming Palace National
Heritage Site Park, Xi'an
(2007)

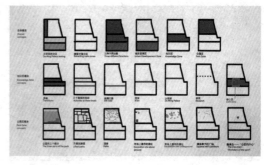

图 .10-4
挪威 "TEAM
3+" 提案的分区
原则

（图源：西安大明宫国家遗址
公园概念设计国际竞赛挪威团
队参赛方案，2007年）

Figure.10-4
Zoning principle

(Photo credit: Proposal
from Norwegian Team 3+
for international conception
design competition on Xi'an
Daming Palace National
Heritage Site Park, Xi'an,
2007)

图 .10-5
2007年的大明
宫遗址南区
（西安，2007年）

Figure.10-5
Existing
housing on
the Daming
Palace site
Xi'an (2007)

图 .10-6
拆除后的一片
" 白板 "
（西安，2012年）

Figure.10-6
Tabula rasa
Xi'an (2012)

图 .10-7
今日南区
（西安，2014年）

Figure.10-7
The southern
zone today
Xi'an (2014)

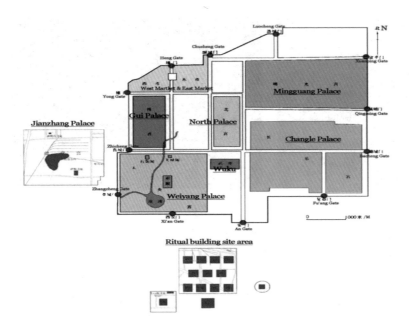

图 .10-8
汉长安城宫殿区
域分布图

（图源：汉长安遗址保护中挪
联合研究报告基础资料汇编，
2014年）

Figure.10-8
Palace district
of the Han
Chang'an City
site

(Photo credit: Case Han
Chang'an City, Project
Description, Xi'an, 2014)

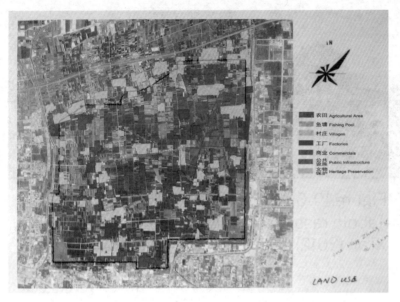

图 .10-9
汉长安城遗址区
域2010年用地
使用状况分析图

（图源：汉长安遗址保护中挪
联合研究报 告基础资料汇编，
2014年）

Figure.10-9
Land use 2010-
illustrating
the composite
situation

(Photo credit: Case Han
Chang'an City, Project
Description, Xi'an, 2014)

图 .10-10
调查团队与学校
的孩子们聊天
（西安，2010年）

Figure.10-10
Chat with
school
children
Xi'an (2010)

图 .10-11
即兴电影放映
（西安，2010年）

Figure.10-11
Improvised
movie
performance
Xi'an (2010)

图 .10-12
小学生所绘未央
宫遗址
（西安，2010年）

Figure.10-12
Pupil's
drawing of
Weiyang
Palace
Xi'an (2010)

图 .10-13/14/15
村中生活场景
（西安，2010年）

Figure.
10-13/14/15
Village life
Xi'an (2010)

图.10-16/17
村庄调研
访谈即景
（西安，2012年）

Figure.10-16/17
Interview
situations
Xi'an (2012)

图.10-18
2012年12月
大刘村村庄状况

Figure.10-18
Daliu Village
in December
2012, Xi'an

图.10-19
2013年3月
大刘村村庄状况

Figure.10-19
Daliu Village
in March 2013,
Xi'an

图 .10-20
2014年9月大刘
村村庄原址状况

Figure.10-20
Site of Daliu
Village in
September
2014, Xi'an

图 .10-21
拆迁安置前的
大刘村社交生活
场景
（西安，2013年）

Figure.10-21
Social life
in Daliu
Village before
resettlement
Xi'an (2013)

图 .10-22
村民安置
住宅区介绍
（西安大刘村，
2013年）

Figure.10-22
Presentation
of resettled
housing
district
Xi'an (2013)

The Indian, non-violent freedom fighter Mahatma Gandhi once said: "I do not believe in the doctrine of the greatest good of the greatest number. It means in its nakedness that in order to achieve the supposed good of fifty-one percent the interest of forty-nine percent may be, or rather, should be sacrificed. It is a heartless doctrine and has done harm to humanity. The only real, dignified human doctrine is the greatest good for all." * 5

The uncompromising attitude of Mahatma Gandhi's ideal is touching. I believe everyone would like to live according to Gandhi's principle, but many people might also find it a bit of a naive ideal and certainly very difficult to practice. As soon as a great number of people are engaged in a process, there will be contradicting needs and dreams. So how are we to make the final decisions? How are we to know the variety of opinions and priorities of the involved persons, the so-called stakeholders? There will always be decision-makers of some kind at the end of the process. What kind of knowledge do they have regarding what people want, how they get it, and how do they relate to it? This chapter will reflect on these questions within the field of conservation and development of historic districts, first by looking at international and national doctrines, then by describing how these questions are discussed in two cases where my university has been involved. The one is an international competition for the development of an archaeological park, on the site of the Daming palace in Xi'an. The other is a research project on Han Chang'an City, the Han Dynasty Weiyang Palace area, also in Xi'an.

The scope of interest in cultural heritage in China, both when it comes to what should be considered a cultural heri-

---

* 5   Quotation from Gandhi Ashram Museum in Ahmedabad, India.

tage and how to deal with it, has, by and large, followed the international evolution of doctrines. The main difference between the national and the international practice seems to be the involvement of common people in the processes of conserving and developing the sites. This difference becomes essential when international rules meet national legislation where projects from China are to be evaluated according to international regulations, for example in applications to admit sites into UNESCO's World Heritage List (WHL). However, in the document named "Principles for the Conservation of Heritage Sites in China," in short named "China Principles" (originally dated the year 2000, and revised in 2015) * 6, article 8 on participation says "Conservation of heritage sites is a social undertaking that requires broad community participation. The public should derive social benefit from heritage conservation."

The revised China Principles show that the national regulations follow the development of international revisions. The burning question then becomes how the regulations are interpreted, obeyed, and practiced. The conservation and development of inhabited historic districts will always be full of compromises. Recommendations describe ideal conditions and best practices. As in all other countries, there are gaps between theory and practice. Some adaptations may be necessary, others not. In the following, I will elaborate on some issues mentioned above.

---

* 6   A document worked out by the State Administration of Cultural Heritage, in cooperation with the Getty Conservation Institute, to collaboratively develop national guidelines for cultural heritage conservation and management that respect and reflect Chinese traditions and approaches to conservation.

THE DILEMMA
OF BEAUTY
AND TRUTH

One dilemma in the protection of cultural heritage is the aesthetical aspect. In this light, the main questions concern the basic intention of the conservation: Is the main aim of the conservation to beautify the object, or is it more important to let the repair show the tear and wear of time? This has been discussed by conservationists and architects for ages. The Venice Charter * 7 as well as recent doctrinal texts, like the revised China Principles, recommends that new additions to a cultural heritage object should be conspicuous, and there are many who consider this principle problematic to the aesthetical harmony of the object.

We always have an idealized image of the object in question, and our image is shaped by our idea of the object's state at the time of its creation. This idea of the object's original state can, of course, be based on flawed and uncritically information. The dilemma also occurs when we employ a conventional understanding of what is beautiful and what is not. Being beautiful can then often mean smoothing out all irregularities that we find disturbing. For our understanding of the past, it seems to be necessary to modify our aesthetical ideals, to also encompass the footprints of historical change, similar to how we might find beauty in the wrinkles of an old, human face.

＊7 The Venice Charter for the Conservation and Restoration of Monuments and Sites is a set of guidelines, drawn up in 1964 by a group of conservation professionals in Venice, that provides an international framework for the conservation and restoration of historic buildings. Adopted by ICOMOS in 1965.

In those complicated matters, two principal attitudes of conservation unfold. One is to remove anything but the features directly referring to the initial phase of the site's creation. The other contrasting principle is to consider everything that has happened in the aftermath of the current era as an important part of the site's history. The consequence of this last principle is to show the specific period of time, the time of the site's original creation, as part of a continuity of historical events. In other words, it shows that the site's history starts before the time of its original creation of the site and moves up through time to the present day. History lingers and all periods leave their trace.

The choice of principle has substantial consequences. In both the Daming Palace and the Han Chang'an case (see below), the first principle was followed, and as a result, the local villagers were moved out and resettled somewhere else.

<u>THE DILEMMA</u>
<u>OF BUREAUCRACY</u>
<u>VERSUS CLIENTELISM</u>

In social sciences we distinguish between two kinds of governance: Bureaucracy, which means equal rights for everyone according to directives of law; and Clientelism (often named a "Patron-client" system) where the patron gains the support or the solidarity of the client, and the client gains an advantage in exchange. Rule-of-law is bureaucratic, while clientelism is more flexible. Clientelism implies possibilities for misuse of power, but also possibilities for a more flexible adaptation to the concrete reality of a situation.

The Chinese phenomenon *Guanxi* may be considered a special version of clientelism, but rather than a one-to-one relationship, it is a network constellation consisting of several persons. As I have understood it, there are mutual obligations within the network. It is flexible but regulated by unwritten agreements and a common understanding of what to do and not to do. The only direct experience of Guanxi I have had, a very positive one, was during our first visit to China. We met a university colleague in Beijing, and when he heard that we intended to make a tour of the country, he gave me a list of university classmates with their phone numbers and information on where they were living and working. He told us to just contact them when necessary, with compliments from him, and ask for assistance should we be in need. We did get in touch with some of the people on this list, and among other things, it resulted in contact in Xi'an, which developed into the starting point for engagements over the next 30 years.

A classic dilemma when planning and implementing plans - and this is very much the case in China - is the problem of numbers. When there is a great number of families affected by interventions, direct communication with each family takes up many resources. A bureaucratic model of governance, which is normally the preferred one in China, as it is in many other countries, will often reduce contradictions and time-demanding debates. Taking each family's needs and wishes into consideration could easily create jealousy between families, or - when changes are effectuated - give the families a feeling of being treated unjust and unequally (like, for example, in cases of expropriation and resettlement). * 8 And when the aim is to create a har-

---

* 8   Underlined in a personal conversation with the leader of the master planning of the Han Chang'an City site.

monious neighborhood, stirring up such feelings would obviously defy the purpose. On the other hand, by taking individual conditions into consideration, the premises for the planning may change substantially, and other ways of solving problems may reveal themselves. For instance, in order to efficiently organize an urban renewal project, it would be smart to first investigate who would prefer to stay and who would prefer to move out. At the end of the day, this strategy might save both time and other resources. * 9

Let us, with these reflections in mind, look at the cases of the Daming Palace and the Han Chang'an sites.

THE DAMING PALACE
COMPETITION 2007

The palace area of the Tang Dynasty capital Chang'an, located to the north of the old Xi'an railway station, was a site encompassing 3.6 square kilometers with farmland, small-scale industries, housing areas, important historical relics, and palace architectural ruins. Being on the first batch of national cultural relics in 1961, and being a most representative of the Tang civilization, a heyday of imperial China, the time was ripe for interventions to establish a national archaeologic park. To do so, the first stage was to find ideas, which could give guidelines for the future planning of the area. Within this framework, Xi'an Municipal Bureau of Cultural Heritage and Xi'an Qujiang Preservation and Renovation Office of Daming Palace Ruin Area organized an international competition in October 2007.

---

*9   To make this kind of clientelism a constructive option, it would be necessary to closely control the power of the patron.

After a prequalification round, eight teams were selected to take part in the final review. One of them was based in Trondheim, consisting of architects and planners from NTNU, the architect office BARK, and the landscape and architect office AGRAFF. For a period of time, the team was also joined by a Chinese colleague from Tsing Hua University. This configuration gave a name to the team: Team 3+.

The basic priorities for our project were to reduce environmental stress (on traffic systems, green areas, architectural structures); careful management of the archaeological structures (minimal digging, lightweight constructions on plate foundations, protection of existing archaeological formations); maintain existent habitations for the present residents; avoiding conjectural "copies" of unknown structures of the past.

Our main master plan elements referred to the original zones of the Tang Dynasty Palace. To the south was the most public area in the Tang Dynasty palace. The project proposed an urban development zone there where the most valuable parts of the building volumes were maintained and developed, and when adding new buildings rules should be applied to control building heights. The middle zone was used for the formal administration in the Tang Dynasty period – in our project we proposed that it should be a knowledge zone with museums, educational facilities, and exhibitions of the archaeological excavation material. The northern zone was used for informal activities and the family life of the emperor – in our project, we proposed it should be used as a public park.

We chose to give attention to the southern zone with the 20th-century buildings and streets. Some buildings here were in a very shabby condition, others very good, most of them somewhere in-between. These buildings were valuable in three respects: They were the homes of people who had resided here for generations – some of these were families of former migrants from Henan Province who still spoke their original home dialect; they were a representation of important periods of the Chinese history; and finally, as a contrast to the monumental scale of the Tang Dynasty palace, they could serve to illustrate the relationship between daily life structures and the vast dimensions of the ruling class buildings. We recommended making detailed studies of the existing buildings and evaluating them. Our suggested method for this was to go into dialogues with people residing in the area, particularly in the southern zone, and find out what was worth restoring and what could be removed – rather than a wholesale demolishing of the whole district.

The strategy chosen in the end was different: When realizing the project, demolishing all buildings in the southern zone to give space for the archaeological park.

In June 2014, the first section of the Silk Road, the so-called Chang'an-Tian-shan Silk Road Corridor, was listed on UNESCO's WHL. Xi'an, as an Eastern terminal, would be represented by the Weiyang Palace site, one of several palaces in the whole Han Chang'an Palace district shown on the map below.

Located just outside the second ring road, to the Northwest of Xi'an city center, and covering an area of all together 36.5 square kilometers, the Han Chang'an City site represented an urban area under hard pressure from urban expansion and development. It was inhabited by 60-70,000 people in 54 villages, many of whom were farmers who had land inside the palace area. There were regulations on farming, restricting deep digging, in order to avoid ruining the many subterranean, cultural relics. This regulation affected the economy of the farmer families. Starting in 1959 there had been archaeological excavations on the site, with important findings casting light on the life of the Western Han Dynasty (206 BC-AD 5). However, most of the ruins were under the surface and still not dug out. Like the Tang Dynasty, the Han Dynasty is considered to be a peak period in Chinese history, so on the local level as well as on the national level, the site represents a most valuable identity marker. The expanding tourist business also paid much interest to the site.

This situation was complex and difficult to manage. Political discussions had been going on for decades. And now a new perspective was introduced, adding another layer to the discussions: If you want to get on the WHL, it is hard to circumcise the international eye of UNESCO, with its regulations and conventions. The Silk Road project * 10 was multi-national, which complicated the procedures, time schedules, and processes for the involved countries. On strict deadlines, decisions had to be taken in order to hand in the appropriate material for the application.

Sponsored by the Norwegian embassy in Beijing, researchers from NTNU and XAUAT joined forces. Over a period of three years, between 2011 and 2014, they were to carry out a program, entitled "Involvement of Local Stakeholders. Development and Conservation of Han Chang'an City Historic Site." The team used several methods for studying the case: Regularly following what was said in the media (newspapers, TV, radio, internet); interviews with stake-

---

*10   A so-called "Serial World Heritage" in the UNESCO system.

holders (villagers, local businessmen/women, local leaders, planners, archaeologists, historians); personal observations during site visits; and seminars with a reference group of central decision-makers, professionals, and planners in the project.

There are many ways stakeholders may be involved. Here is an example. The year before the research project on Han Chang'an City started, teachers from NTNU, in cooperation with XAUAT, arranged a master course on the site, investigating the potential of the situation and employing several methods of involvement. Besides making design proposals, the students also organized activities for getting in touch with the villagers. For instance, they arranged a film session together with a local resident who in old days used to show films in the evening. He still had his old film projector and a pile of old films. A wall was painted white to make a screen. In the dark evening, the movie screening attracted most villagers. Another team borrowed a class of schoolchildren from the village school and organized a treasure hunt. The pupils were to find cultural relics hidden on the site of important Han Dynasty buildings. They also made drawings to be shown on an outdoor exhibition terrace.

An eleven years old schoolgirl from Da Liu Zhai village wrote:

I like running in the fields and playing together with my friends. In the fields, there are a lot of mounds, and we are usually crawling on top of it happily. Then suddenly one day, a mound in the west of our village was covered with wood. My mother told me that the mounds around the village are all part of Han Chang'an city. Though they are ruins now, they are still a treasure. So amazing! Several of our good friends at once went to see the protected mounds. It looks beautiful, and we were lying on the wood and felt very good. I hope all the ruins here can be protected. And then our village will be more beautiful–ah!

Why this interest in the involvement of local stakeholders? The international doctrinal texts from the United Nations, UNESCO, and ICOMOS emphasize-with reference to Human Rights - the importance of this aspect. For example: Applying for a place on WHL, one has to document how local people have been involved and informed, and how this involvement will be maintained in the future development and management of the site. UNESCO also emphasizes this aspect in The World Heritage Convention Operational Guidelines.

We used the international, doctrinal texts on this matter as benchmarks when we observed and assessed how different, local stakeholders were involved and how they were informed – when and by whom. Two of the villages in the Weiyang area, Daliu village, and Gaomiao village, were selected for the studies. Villagers were regularly interviewed every second month of the project period. We were cordially received everywhere. When private property and future life are at stake, such situations will always be, anywhere in the world, sensitive. And because the situation was so fragile, we had some restrictions, especially in the first year, on which topics we might introduce. Later we felt free to pose any questions. Some interviewees expressed themselves very carefully, but most people we talked with liked to talk about their daily life, how they conceived the situation, and which expectations they had for the future.

Between December 2012 and February 2013, approximately 15,000 villagers had been relocated and their 9 villages demolished. In autumn 2012, just before relocation, the residents received solid briefings. Before that, very little information was given by the authorities to the villagers. What they knew was based on rumors, verbal ones, or on the internet. It was not possible for them to know what the destiny of their village could be, and thus it was hard to plan for their future.

Only politicians and professionals were in truth involved in the project. Local villagers and businessmen/women were not involved in other ways than in practical matters when they had to leave their homes and remove their ancestors' tombs. Feelings around being relocated to a new high-rise district outside the third ring road varied from generation to generation. Generally, old people would have preferred to stay in their old street where they could have a chat with neighbors. Young people were looking forward to a change and new possibilities. The young people's attachment to place was, in other words, different and because of the uncertainties about their future maybe their sense of belonging to the location faded away during the long process.

What I have tried to show in this chapter is that by selecting the strategy of removing all traces of historical footprints other than those of the Tang Dynasty in the Daming palace site, and those of the Han Dynasty in the Han Chang'an City site, the consequence was that the involvement of local stakeholders was minimized.

11

农村生活

11

LIFE IN COUNTRY-SIDE

城市化极大地改变了中国社会。1985年我初次到访中国时，当时的总人口是10.5亿，生活在农村的人口约占75%。今天，14亿人口中约有35%~40%生活在农村，目下仍有4亿~5亿的农村人口。在快速发展扩张的城市和缩小的乡村，许多方面的变化都引人注目。如此迅速地迁入城市，意味着城市新移民和他们成长的乡村之间有着情感和现实的联系，他们骨血里仍有乡村生活的印迹。每当我看到巨大城市群中新建的高层居住区，念及数以亿计的人口从农村迁往大都市，总是在想：他们如何适应新环境？他们对离开家乡作何感想？他们与以前的存在有着怎样的联系？如果能在几个高楼大厦中做详细的统计调查，找出生活在那里的人，了解他们现今与过往生活相比如何，这会是一个很棒的项目。

乡村意味着粮食生产。通过对技术和农业资源的选择，景观地貌、土壤特性、水路河道被农业改造和利用。在不同文化地景下，人们找到生存之道，其生活节奏取决于农耕条件。他们在与自然的紧密互动中构建自己的环境，组织物质干预。参观农田，可以很好地了解中国民族性格和历史的一个基本深层结构。以下是我作为一个外国人，觉得既美好又有意义的乡村活动集萃。

玉米在很多国家是基本食物原料，在中国也如此。收获的季节是风景如画的时分。村里到处是黄澄澄的玉米棒和穗子，像大串香蕉一样四处悬挂着，使村庄街巷院落平添色彩。剥玉米棒也是当地人们的社交活动。废弃物被收集起来，用于其他用途，如喂猪和用作炉中燃料。玉米（粒）放在任何有硬化表面的地方晒干。

看着农村的玉米收成，人会有感于大自然的丰饶，觉得在这个常令人疑虑困惑的变化发展时代，世界毕竟不是那么糟糕。

除了利用自然生产粮食，还有很多方式将自然资源用于其他用途。即使在工业化时代，仍然可以找到利用当地资源进行小规模（手工）生产的有趣实例。其中一个例子即陕西柞水凤凰镇的造纸业。这里利用构树的树皮纤维生产高质量的纸张。这种纸有不同用途，如在下葬前包裹亡者。*19但更有趣的是，它也被艺术家们使用，他们可以在城里昂贵的艺术文具商店中买到这种纸。另一个例子是很多铁匠从凤凰镇周边山上取铁矿打造农具。

窑居非常适合黄土地貌。这里土质松软，一锹下去即能成型，而土质又足够坚硬形成稳定的垂直壁面。据说一个成年人带着其长子用不到一周时间就能用铁锹掘出一孔窑洞。窑居有显而易见的优点：造价低廉且四季宜人。（窑内）空间被周围土层加热，全年保持恒定温度，冬暖夏凉。即使今天，黄土高原上仍有很多家庭住窑洞。老人们更喜欢窑居，觉得较通常的水泥房更为舒适。而年轻人则觉得他们落伍，想要过更现代的生活。

这些窑洞大致有两类，这两种类型在任何（有窑洞的）地方都能同时见到：靠崖式窑洞和下沉式窑洞（地坑窑），后者的形式是向下挖出庭院大小的井院。亦有将窑居现代化的尝试，例如加设通风井使空气流通。我所见另一类现代化改造是一种两层窑洞，这样可以有更丰富的外立面，以及更多功能独立的房间。

20世纪70年代，"农民画"在欧洲毛派中很流行。据说，普通农民在午饭时离开田间地头，在画架前画上片刻。然后他们回到田里，直到傍晚一天劳作结束后，才继续作画。据说这是任何中国农民掌握的技能。后来我们才知道，这些画家其实是职业画家，所作都是精品。在陕西户县（今鄠邑区）有一个农民画博物馆。参观博物馆时，没想到展出的画作是出售的，也就是说，一个普通游客可以从博物馆里把画作当下买走（我得坦白自己是这样做的）。后来我知道，中国百姓围绕着"原作"和"临摹"的概念，与西方人对此的典型想法有些不同。有人告诉我，一幅好的复制品，可能和原件一样有价值。

乡村文化有很多有趣的表现形式。以下是我们听到的另一个例子：传说在唐朝时，宫中有两派势力。某次其中一方为保命不得不逃

---

*19　译者注：入殓时以黄纸裹身为我国多地葬俗，覆于亡者身上纸张谓"阴阳纸"，多以轻薄黄麻纸或白土纸为最佳。

亡。逃难者中有一个乐班，带着乐器乐谱。他们在都城外的一个村庄定居。乐班的音乐从此得以留存，至今仍是村里一个活的传统——村里一个业余班子仍在演奏着据称源于唐代的西安鼓乐。

近几十年来，中国快速的城市化和现代化进程无疑意味着各种文化习俗的变化。以婚庆仪式为例：庆祝活动本身，以及即将结婚的夫妇如何在婚前见面约会，双方的家庭如何参与。这些过程都随着时代社会发展而变化，并在城市和农村生活相互影响中形成差异和相似之处。所谓"包办婚姻"依然存在，但现在准新娘、新郎通常有最后的决定权。今天，婚姻通常是建立在一段时间的约会而相互爱慕基础上的。

约会本身也发生了变化，这既涉及年轻人间的风气习俗，也涉及公共道德认为体面的规范。换而言之，即约会的公开化程度。城市中大学的情况，可以作为一个随着时间推移文化转型的例证。在大学里，过去禁止学生公开显示与他人的亲密恋爱关系。直到20世纪90年代初，甚至仍禁止牵手。男生到女生宿舍会受到惩处。如今，情侣们紧贴着走在一起、坐于一处，在大庭广众之下亲昵。诚然，这种情况仍然不像有些国家那样普遍，但已经普遍到足以说明关于何为得体的实质性改变。换句话说，即文化的改变。

在城里，有专门的（婚庆）公司来安排婚礼：请柬、聚会、拍照，以及最重要的，新娘的婚纱礼服和新郎的西装。在农村，一些传统习俗仍然存在，但更多是非正式的做法，全村参与其中，无论帮忙准备还是参与庆祝。城乡两地的婚礼可能不乏相同的仪式——致辞、敬酒、送礼、嘉宾、游戏抢亲、在新娘家"破"门而入，等等。另一方面，很多内容则在发生实质性的变化。本章中的一些图片，显示了乡村和城市婚庆风格的异同。

1989年，我骑车游走于昆明周边的乡间，看着农民牵着水牛以相当古老的方式在田里耕作。我们路过一位坐在路边抽着烟斗的男人，他向我们招招手，想和我们聊聊。令我们惊讶的是，他竟然会说一点英语，他问我们是哪里人。"挪威。"我答道。"你知道挪威在哪里吗？"我想听听他的回答。他想了想，说道："在北欧，斯堪的纳维亚半岛。挪威是靠近大西洋的国家。"打那以后，我不得不调整自己对别人的知识储备所抱有的偏见。

图 .11-1
剥玉米
（柞水凤凰镇，
2007年）

Figure.11-1
Corn
threshing
Fenghuang
(2007)

图 .11-2
福建泉州渔场
（福建，2012年）

Figure.11-2
Seafood
farming near
Quanzhou
Fujian (2012)

图 .11-3
福建田螺坑村
客家土楼
（福建，2012年）

Figure.11-3
Landscape
of a Hakka
Round Houses
settlement,
Tianluokeng
Tulou Cluster
Fujian (2012)

图 .11-4
福建田螺坑村
农人
（福建，2012年）

Figure.11-4
A villager of
Tianluokeng
Tulou Cluster
Fujian (2012)

图.11-5
收集秸秆
(陕西，2012年)

Figure.11-5
Collecting
wastes
Shaanxi (2012)

图.11-6
晾晒玉米
(柞水凤凰镇，
2007年)

Figure.11-6
Drying corn
Fenghuang
(2007)

11 农村生活

图 .11-7/8/9
收获季节
（陕西，2012年）

Figure.
11-7/8/9
Village life in
harvesting

season
Shaanxi (2012)

图 .11-10
边干活边聊天
（柞水凤凰镇,
2007年）

Figure.11-10
Working and
chatting
Fenghuang
(2007)

图 .11-11
劳碌一天后
的安详时刻
（陕西, 2007年）

Figure.11-11
Peaceful at
the end of the
day
Shaanxi (2007)

图 .11-12
食物，一切
生活的基础
（柞水凤凰镇,
2008年）

Figure.11-12
Food–the very
basis of life
Fenghuang
(2008)

11 农村生活

图 .11-13/14
传统手工造纸
（柞水凤凰镇，
2005年）

Figure.11-13/14
Traditional
craft of hand-
making paper
Fenghuang
(2005)

图 .11-15/16
铁匠铺
（柞水凤凰镇，
2007年）

Figure.11-15/16
The
blacksmith
Fenghuang
(2007)

图 .11-17
黄土台塬景观
（陕西永寿县
等驾坡村，
2007年）

Figure.11-17
Loess
landscape near
Dengjiapo
Village
Yongshou,
Shaanxi (2007)

图 .11-18 山坡窑洞 （延安，1989年）	Figure.11-18 Caves in a slope Yan'an (1989)	图 .11-19 普通房屋与窑洞 的混合 （延安，1989年）	Figure.11-19 A mix of regular and elaborated caves Yan'an (1989)

图 .11-20	Figure.11-20	图 .11-21	Figure.11-21

图.11-20/21/22
地坑窑
（永寿县
等驾坡村，
2007年）

Figure.
11.20/21/22
Sunken
courtyard
cave
dwellings,
Dengjiapo
Village
Shaanxi (2007)

图.11-23/24
陕北窑洞土炕，
坐卧皆宜
（延安，1989年）

Figure.11-23/24
Kang–good
for sleeping
and working
Yan'an (1989)

图 .11-25
现代窑洞
（陕西，2000年）

Figure.11.25
Modern caves
Shaanxi (2000)

图 .11-26
"农民画"画家
和他的作品
（陕西，1991年）

Figure.11-26
The artist
of "farmer
painting" in
front of his
work
Shaanxi (1991)

11 农村生活

图 .11-27/28
乡村班子演奏
西安鼓乐
（西安市长安
区何家营村，
2007年）

Figure.
11-27/28
Village
orchestra
playing the
Tang Dynasty

music,
Hejiaying
Village
Xi'an (2007)

图 .11-29
全村人参加婚礼
（西安市长安区
子午镇张村，
2009年）

Figure.11-29
The whole
village
taking part,
Zhangcun
Village of
Ziwu Town
Xi'an (2009)

图 .11-30
临时搭建的厨房（西安市长安区子午镇张村，2009年）

Figure.11-30
Improvised kitchen for the feast, Ziwu Town Xi'an (2009)

图 .11-31
记录红包礼金（西安市长安区子午镇张村，2009年）

Figure.11-31
Making notes on who gave gifts, what and how much Ziwu Town, Xi'an (2009)

图 .11-32
新婚夫妇和专业主持（西安市长安区子午镇张村，2009年）

Figure.11-32
The couple and the professional toastmaster Ziwu Town, Xi'an (2009)

图 .11-33
期待（西安，2011年）

Figure.11-33
Expectations Xi'an (2011)

11 农村生活

图 .11-34
孔庙婚纱留影
（西安，2019年）

Figure.11-34
Wedding
photos at a
Confucian
temple
Xi'an (2019)

图 .11-35
国外旅行结婚
（意大利佛罗伦
萨，2014年）

Figure.11.35
Traveling
abroad for
marriage
Florence, Italy
(2014)

图 .11-36
乡间
（昆明，1989年）

Figure.11-36
Countryside
Kunming
(1989)

图 .11-37
抽水烟的男人
（昆明，1989年）

Figure.11-37
The man who
smokes water
pipe
Kunming
(1989)

Urbanization has drastically changed Chinese society. During my first visit to China in 1985, approximately 75 percent of the population (at that time the population counted 1.05 billion people) lived in rural areas. Today, approximately 35-40 percent of 1.4 billion people live in the countryside. In many ways, the changes have been conspicuous in fast-growing cities and in shrinking villages. But still, there is a remaining rural population of 400-500 million people. Such a rapid migration to the cities means that the new urban citizens have emotional and practical ties to the villages where they grew up – they have rural life in their backbones. When I see the new skyscraper housing areas in huge agglomerations, and I think of the hundreds of millions that have moved from the rural villages to the metropolises, I always wonder: How do they adapt to the new situation? How do they feel about leaving their native village? How are their ties to the former existence? It would be a fantastic project to do detailed and statistical research on a couple of skyscrapers, in an attempt to find out who lives there and how their present lives compare with their former lives.

Countryside means food production. The landscape's topography, soil properties, and waterways are transformed and utilized by farming, a selection of technology, and by agricultural resources. In the various cultural landscapes, people find their ways of surviving, their rhythm of life dictated by farming conditions. They construct their environment and organize their physical interventions in close interplay with nature. Visiting the farmland areas gives a good insight into a basic and deep, structural part of the national character and the history of China. As a reminder of this, here follows a selection of countryside activities that I, as a foreigner, found both beautiful and meaningful.

Corn is a basic food ingredient in many countries, and in China as well. The harvesting season is a picturesque time. The villages are full of yellow ears and cobs, hanging everywhere like banana clusters, coloring village streets, and courtyards. Peeling the corncobs gives possibilities for social interaction. Waste products are collected and used for other purposes, like feeding the pigs and fuel in ovens. Corn is dried anywhere there is a hard surface.

Looking at the corn harvest in the countryside, one gets an impression of bountiful nature, and a feeling that, in times of changes and often dubious development, the world is, after all, not so bad.

Besides using nature for food production, there are numerous ways of utilizing natural resources for other purposes. Even in our industrialized age, one can still find interesting examples of small-scale production based on local resources. One example is papermaking in Fenghuang of the Shaanxi Province. Here the tree bark fibers from the paper mulberry (in Chinese sounds like "Dog-skin tree") are utilized to produce a high-quality paper. The paper is used for different purposes. It is, for instance, used to wrap bodies before burial. But more titillating, perhaps, it is used by artists who can buy this paper in expensive art shops in the cities. Another example is the many blacksmiths who make farming tools from the iron of nearby mines located in the mountains surrounding the town.

The loess landscape is perfect for housing in the form of cave dwellings. The soil is soft enough to be formed by a spade and hard enough to form stable, vertical walls. It is said that a man with his oldest son can dig a cave with spades within one week. The cave dwellings have obvious advantages: They are cheap and they are comfortable in all seasons. The space is heated by the surrounding earth and keeps a stable temperature during the whole year, warm in winter, and cool in summer. Even today, there are many families living in caves on the loess plateau. Old people prefer caves, finding them more comfortable than what is often the alternative, a concrete building. Young people find them old-fashioned and want a more modern life.

There are basically two types of these cave dwellings, and they are found anywhere: Caves in slopes and caves on horizontal ground, the latter in the form of a courtyard-sized well dug downwards. There are attempts at modernizing the cave dwellings, for instance by adding ventilation shafts in order to create air circulation. Another modernization I have come across is a two-story cave (see the illustration), which allows for more façade area and the possibility to have additional separate rooms.

So-called "farmer paintings" were popular among European supporters of Maoism in the 1970s. The story goes that the common farmers left the fields at lunchtime, running to the easels to do a few moments' worths of painting. Then they would return to the fields, only to continue paintings in the evening, after the workday was over. Supposedly, this was a skill any farmer had. It was revealed later that the artists were indeed professional painters. One way or the other, the paintings were fine. In Huxian County, there is a farmer painting museum. Visiting the museum, it was a surprise that paintings on display were for sale and could thus be immediately removed from the museum by a casual visitor, (which I have to admit, I did). Later I learned that the common people's ideas around the concept of "original" and "copy" are a bit different from the typical Westerner's idea. A good copy, I was told, maybe as valuable as the original.

There are many interesting expressions of rural culture. Here is another one: In the Tang Dynasty, there were two power fractions in the imperial palace. At some point, one of them had to flee to save their lives. Among the refugees was an orchestra, bringing their instruments and music notations. They settled in a village some miles outside the city. The music of that orchestra has been protected since then and is still a living tradition in the village – nowadays there is an amateur orchestra in the village playing the Tang Dynasty music.

The rapid urbanization and general modernization in China during the recent decades, imply, of course, changes in a wide specter of cultural phenomena. Take marriage rites, par example; the celebration itself and also how the couple to be married meet and date before marriage, and how families on both sides are involved. These processes are flavored by the societal development in time as well as differences and similarities evolved in the mutual influence between urban and rural life. Arranged marriages still take place, but now the future bride and groom normally have the last word. Today, wedlock is normally based on falling in love and a period of dating.

Dating itself has also changed, both when it comes to customs among young people and to norms for what public morals find decent. In other words, how openly dating is practiced. An example from the university sphere (which is urban) could be mentioned as an illustration of cultural transformation over time: In the universities, it used to be forbidden for students publicly to display amorous relationships with another person. As late as the early 1990s, it was even forbidden to hold hands. Visits to the female dormitories by a boy were penalized. Today couples walk and sit tightly together, flirting in public view. Admittedly this might still not be as common as it may be in some other countries, but common enough to demonstrate a substantial alteration in what is comme il faut. Cultural change, in other words.

In cities weddings down to all details are managed by special firms: Invitations, parties, photo sessions, and last but not least, the gown for the bride and suit for the groom. In the countryside, traditional customs are maintained, but practiced more informally, with the whole village preparing and taking part in the celebration. The same rituals may be carried out in both places – speeches, toasts, gift giving, guests, humorous kidnapping of the bride, "breaking" the gate of the home of the bride, and so on. On the other hand, much is undergoing substantial changes. Here follows some pictures illustrating the differences and similarities in the style of village and city weddings.

Finishing up this rural cavalcade: In 1989, I was in Kunming, biking around in the surrounding countryside, watching the water buffaloes and the farmers plowing the fields in a rather old-fashioned way. We passed a man sitting by the roadside smoking his pipe and he waved us over to talk with us. To our surprise, he spoke some English, and he asked us where we were from. "Norway," I answered. "Do you know where Norway is?" I was curious to hear his answer. He thought for a while, then said: "It is in Northern Europe, in Scandinavia. Norway is the country close to the Atlantic Ocean." After that, I had to adjust some of my prejudices on who knows what.

12

有关
风水的
村庄
凤凰
2009

12

A VILLAGE
OF *FENG
SHUI*
Fenghuang
2009

自古以来，风水对于（中国）乡村的本土化和组织布局起着决定性的作用。一度，看风水被法律禁止，今天，在现代社会风水影响日衰。虽然大多数人都知道它的历史和某些特征，但越来越少的人真正相信它的价值。风水的原则与建筑和周边自然环境的关系密切相关，赋予了自然元素（地形、河流、拓扑结构、自然形态）在人工环境中的特殊意义和角色。因此，即使对这一理论并非信从，当人们欣赏赞叹周边的美景和街巷、庭院、建筑的组织，并为之自豪时，也会对这一与自然息息相关的文化心有所感。在这方面，风水代表了建筑与自然紧密联系的典范。

凤凰古镇位于秦岭南麓的柞水县。这是一座有着历史重要性的小镇，部分出于其作为当地行政中心的历史角色，部分是由于其关键性的地理位置。三条河流交汇于此，其中一条直接与长江水系相通，是连接中国北方与南方的交通要道。向北越过群山，一条古道连接着凤凰和渭河流域的肥沃平原。因为这些因素，凤凰自古就是一个重要集市，南来北往的商品货物在此交换。村中有一条保存完好的老街，长约500米，街道两边是传统院落和店铺。老街蜿蜒盘转，形如鲤鱼，据说是出于风水的考虑。其地理选址也是如此，村子三面环山，避开北面的寒风，向南边波澜起伏的河面敞开。

对于许多家庭来说，今天仅靠种地已然难以维持生计。村子面临家庭外迁的问题，为避免人口减少，他们试图通过旅游业发展来应对这一挑战，这是中国许多乡土村落的典型策略。游客被传统的建筑和生活方式、地方手工艺、当地吃食、地方历史、与老院子融为一体的餐馆酒店，以及——对来自大城市的游客来说——干净的空气和优美的自然所吸引。这些促进旅游的努力如果成功，就会创造新的生活方式，同时强化地方认同感。凤凰镇的游客越来越多，使得面朝老街的传统建筑得以很好的维护。这一方面是由于居民对老房子的感情，另一方面则是因为老街是村里的主要旅游景点。

政府部门一度有种倾向，要么忽视旅游产业规划，要么过度采取增加游客量的举措。这是一个很难平衡的问题。当游客人数过多时，古镇最初吸引他们的品质可能就会消失，于是旅游业吃掉了自己的孩子。对于凤凰镇来说，这已然是一种威胁。通往村里的交通已经得到改善，一条新的快速路连接了凤凰镇和柞水县城。在当地人过去洗衣服、与邻居聊天、玩耍的地方，沿河修建了新的步道。现在游客对于公共空间拥有了优先权，由此减少了村民的使用。而当地特色的生活原本是吸引游客的主要因素之一，亦因此减少。成功适应新的大众旅游的一个前提，是当地居民参与规划，而如王燊在其论文中所指出，这在凤凰镇并不存在。

这个问题是全球性的，并非仅在凤凰镇和中国其他传统村落存在。旅游业在世界各地都是一个快速发展的产业。在欧洲，也存在因为大众旅游的规模体量将居民挤出现有物业和公共空间的现象，当地居民对此表现出深切的不满。在张开双臂欢迎游客和控制游客容量之间很难找到合适的平衡点，中国有个成语形容得很形象——"势成骑虎"。

图 .12-1
风土建筑
屋脊装饰
（柞水凤凰镇，
2009年）

Figure.12-1
Roof ridge
decoration of
local Vernacular
building
Fenghuang
(2009)

图 .12-2
小镇的日常生活
（柞水凤凰镇，
2005年）

Figure.12-2
Daily life
Fenghuang
(2005)

图 .12-3
农家枕箱，内存
贵重物品，以策
夜间安全。演示
者为当地历史研
究者、中医赤脚
医生（柞水凤凰
镇，2005年）

Figure.12-3
A farmer's pillow
containing
valuables to
be protected
during the night.
Demonstrated
by a local

historian and
barefoot doctor
practicing
traditional
Chinese
medicine
Fenghuang
(2005)

图 .12-4
放学
（柞水凤凰镇，
2009年）

Figure.12-4
After school
Fenghuang
(2009)

图 .12-5
在街边写作业
（柞水凤凰镇，
2009年）

Figure.12-5
Homework on
an old street
Fenghuang
(2009)

图 .12-6
河畔洗衣
（柞水凤凰镇，
2009年）

Figure.12-6
Riverside
laundry
Fenghuang
(2009)

图 .12-7
凤凰镇河畔
旅游设施
（2018年；
摄影：王燚）

Figure.12-7
Riverside
arrangement for
tourists
Fenghuang
(Photo by Wang
Yi, 2018).

Through the ages, *Feng Shui* has played a decisive role in village localization and organization. For some years, practicing *Feng Shui* was strictly forbidden by law; in modern times its influence is reduced. While most people know its history and some of its characteristics, few people really believe in its value. *Feng Shui* links closely to the natural surroundings and gives nature (topography, rivers, topology, natural forms) a special role. So even if the belief in the theory is not strong, when people admire and are proud of the beauty of their surroundings and the organization of streets, courtyards, and buildings, it gives the feeling of culture in deep relationship to nature. In this respect, *Feng Shui* represents an example of architecture closely relating to nature.

Fenghuang is situated in the south of Qinling mountain in Zhashui County. It is a town of historic importance, partly because of its historical role as a political center, and partly because of its geographical location. Three rivers meet there. One of them connects directly to the Yangtze river and is thus a major traffic link to the south of China. Northwards, across the mountains, an old road connected Fenghuang and the fertile plains around the Weihe River basin. Through that, Fenghuang became an important marketplace, where commodities from the north and commodities from the south were exchanged. The village has a well-preserved old street, approximately 500 meters long, with traditional courtyards and shops lining both sides of the street. It is winding in a shape of a carp, which is favorable according to *Feng Shui*. The same is the case for the location, as the village lays sheltered against cold winds from the north by mountains on three sides and opens up towards the undulating river to the south.

For many families, it is now hard to live by farming alone. The village is facing the problem of families moving out, and to avoid depopulation they try to meet that challenge by boosting tourism, a typical strategy for many countryside villages in China. Tourists are attracted by traditional architecture and lifestyle, local handicrafts, local food, the history of the place, restaurants and hotels which are integrated into the old courtyards, and – an important factor for visitors from big cities – clean air and beautiful nature. When successful, these tourism-boosting endeavors create new means of living, and at the same time strengthen the identity of place. As an effect of the growing number of visitors to Fenghuang, the traditional buildings that face the old street are well maintained. This is partly due to the residents' feelings for the old buildings, and partly because the old street is the main tourist attraction of the village.

Governments have a tendency to either neglect planning for tourism or to overdo initiatives put in place to increase the volume of tourists. It is a hard balance. When tourists become too numerous, the qualities that initially attracted them may disappear. Tourism eats its own children. For Fenghuang, this is already a threat. The access to the village has been improved by a new express road linking Fenghuang to Zhashui county town. New walkways along the rivers are constructed in areas where the locals used to wash clothes, chat with neighbors, and play. Now the tourists have hegemony, thus reducing the use by villagers. And, as a consequence, the characteristic local life, which originally was one of the main draws for visitors, is reduced. A premise for a successful adaptation to the new mass tourism is that the local population is involved in the planning, and, as Wang Yi remarks, this has not been the case in Fenghuang.

This problem is global, and not typical just for Fenghuang and other villages in China. Tourism is a rapidly growing industry everywhere. In Europe, there have been manifestations by local citizens expressing deep dissatisfaction with a mass that by its volume pushes residents out of properties and public space. It's difficult to find the right balance between welcoming visitors with open arms and at the same time controlling the volume of tourists. As a Chinese saying goes, "when riding a tiger, it is hard to get off."

13

制陶小镇
陈炉
2007

13

A POTTERY
TOWN
Chenlu
2007

西安向北60公里的耀州窑区，历史上以陶瓷生产闻名，从唐代到元代，这里是中国陶瓷的主产地之一。在耀州传统制陶的几个知名村落中，陈炉自唐代以来即有相当规模的陶瓷生产。在很长一段时间，主要是明、清两朝，陈炉是耀州陶瓷的主要供应地。后来其他乡镇的地位有所上升，但陈炉在整个地区的陶瓷生产中一直占有重要地位。据说，这里窑炉日夜不停地燃烧，因为烟囱中的粼粼烟火，村中永无暗夜。*20 这里家庭窑场众多，竞争激烈，各具特色。自古以来，制瓷技术只传授给家中儿子，因为女儿会嫁到其他家庭，由此泄露生产机密。

随着20世纪50年代的集体化改造，大部分家庭窑场要么关闭，要么被纳入村集体进行陶瓷生产。为了增加产量，村里成立了一个工厂，主要是制作日常生活用品。几年后，又增设一条新的生产线。然而，随着塑料和玻璃器皿越来越受欢迎，工厂面临破产倒闭。工厂欠了一笔巨额银行贷款，这对村里人至今仍然是一个负担（2007年时）。厂里有少数工人仍从事生产，但大部分员工（最多时有1400人）均已下岗，不得不另谋生计。（村里）有几个家庭窑炉作坊仍在工作，向其他地方的商店和游客出售少量陶瓷产品。

从视觉上看，陈炉村的建筑材料以砖为主。当然，除规整的砖墙，最特别的是碎陶器在道路铺砌和各种装饰上的变化运用。最引人注目的是对匣钵星罗云布式的使用。匣钵是一种无盖圆柱体陶制容器，用来保护陶器釉面，使其在燃烧过程中不受火焰、灰土和烟尘影响。匣钵在使用一段时间后，一旦破损，即须从陶瓷生产中剔除。在陈炉，它们被作为建筑材料回收利用。这种做法受到了当地居民的喜欢和游客的赞赏，陈炉也由此具有鲜明的特色和审美表现力。今天，材料的再循环是讨论环境问题的一个重要因素。在陈炉一案中，围绕废旧材料利用的其他问题也应被注意到：再利用的陶瓷元素代表了村子的传统，提醒着居民们，那是他们的家族历史和祖传

---

*20　译者注：据明万历年间《同官县志》记载，"瓷场自麓至巅皆有陶场，土人烧火炼器，炉火昼夜不熄，弥夜皆明，每逢春夜远眺之，荧荧然一鳌山也"。早年间的陈炉村，"瓷场自麓至巅，东西三里，南北绵延五里，炉火昼夜不熄，弥夜皆明，山外远眺，莹莹然一鳌山灯也"。"炉山不夜"因此被列为古"同官八景"之一。参见：刘泽远. 同官县志 [M]. 明万历四十六年刻本。

技艺，从而塑造和强化了村子的身份认同。在砖与陶瓷的"景观"
包围下，人们可以更好地把握陈炉的历史。

陈炉的房屋坐落在景观凹地，四周是农田。整个村子属于上一级行
政单位公社的一部分。在山村幽寂中，声音很容易回响传递。有时，
出殡仪式悲壮尖锐的奏乐声占据主导；有时，则可听到孩子们欢快
的歌声从远处学校敞开的窗户中飘出。这两种声音都凸显了父母一
代或多或少把孩子和爷爷奶奶留在了村里。

当地的学校是一所寄宿制学校，来自其他村子的孩子住在宿舍。我
们遇到的一位年轻老师含着泪水告诉我们，这些孩子很多都太小
了，不应该和亲人分开生活。"他们需要更多的照顾，而这不是我
们能够给他们的……"她说。为了保证自己完成学业后有一份工作，
她申请了这份工作。现在她觉得自己被抛弃了，与城里工作的朋友
和老同学断了联系。校长希望她留下来。她是一个悲伤、认真、孤
独的人。后来重访陈炉时，我们没有见到她，但愿她今天在别处生
活得快乐。

旅游业被视为村里未来经济增长的希望，为了吸引游客，村里在
卫生和建筑标准上做了不少改进。在2017年走访陈炉时，我们可
以看到服务性主街和周边建筑的改善，村边街上新建了几家餐馆、
公厕和一座陶神庙，还有指示牌帮助游客找到还在使用的几座家
庭窑坊。

图 .13-1
陕西陈炉古镇
全景
（铜川陈炉，
2007年）

Figure.13-1
Overall view of
Chenlu Town
Tongchuan,
Shaanxi
(2007)

图 .13-2
制陶工厂
（陈炉，2007年）

Figure.13-2
The factory
Chenlu (2007)

图 .13-3
制陶工人
（陈炉，2007年）

Figure.13-3
Factory
worker
Chenlu (2007)

图 .13-4
家庭窑场陶工
（陈炉，2007年）

Figure.13-4
Potterer –
family kiln
Chenlu (2007)

图 .13-5
家庭作坊
（陈炉，2013年）

Figure.13-5
Family
workshop
Chenlu (2013)

图 .13-6
建筑要素的混合
（陈炉，2007年）

Figure.13-6
Mixed
building
elements
Chenlu (2007)

图 .13-7
院落入口小景
（陈炉，2007年）

Figure.13-7
Entrance
scene
Chenlu (2007)

图.13-8
利用废旧材料
打造乡村景观
（陈炉，2017年）

Figure.13-8
Waste
materials
urban
landscape
Chenlu (2017)

图.13-9
陈炉镇中心小学
（陈炉，2013年）

Figure.13-9
The local
school
Chenlu (2013)

图.13-10
孩子们的宿舍
（陈炉，2008年）

Figure.13-10
Children
dormitory
Chenlu (2008)

图.13-11
学校计算机教室
（陈炉，2008年）

Figure.13-11
Computer
studio at
school
Chenlu (2008)

图 .13-12
爷孙在上学路上
（陈炉，2013年）

Figure.13-12
On the
way to the
school with
grandfather–
not friends
right now·
Chenlu (2013)

图 .13-13
修葺一新
的商业街
（陈炉，2017年）

Figure.13-13
Refurbished
commercial
street
Chenlu (2017)

图 .13-14
山谷中的窑炉
（陈炉，2007年）

Figure.13-14
Kilns in the
valley
Chenlu (2007)

图 .13-15
寒夜过后，
坐在窑炉顶上
暖手的老人
（陈炉，2007年）

Figure.13-15
Sitting on
the top of a
working kiln
warming
hands after a
cold night
Chenlu (2007)

About 60 kilometers north of Xi'an lies Yaozhou district, an area historically well known for its ceramic production. In the period from the Tang (618-906) to the Yuan (1271-1368) dynasties, it was a major producer of pottery in China. Among several villages of the famous Yaozhou pottery tradition, Chenlu has had considerable ceramic production since the Tang Dynasty. For a long period of time, mainly in the Ming (1368-1644) and Qing (1644-1911) dynasties, Chenlu was the major supplier of the Yaozhou ceramics. Later on, other towns were more important, but all the time Chenlu had a major role in the overall regional pottery production. The story goes that kilns were burning day and night, and the village was never dark because of the sparkling smoke from the chimneys. There were numerous competing family kilns, with each having its own specialty. Traditionally the skills were only taught to the sons of the family because the daughters were supposed to marry into other families and could then give away production secrets.

During the collectivization processes in the 1950s, most of the family kilns were either closed down or incorporated into a village collective for ceramic production. A factory was established to increase production, mostly fabricating household objects. A few years later, they added one new production line. Plastic and glass wares were getting more popular, however, and the factory failed and faced bankruptcy. It had a substantial bank loan, which still (in 2007) was a burden for the village. A few workers were still engaged in production in the factory, but most of the staff, which at most counted 1,400 people, was fired and had to find work somewhere else. A couple of family kilns were still active, selling a moderate amount of products to shops elsewhere and to visiting tourists.

Visually the village is dominated by brick materials, regular brick walls, of course, and the most extraordinary thing is the versatile use of broken pottery in paving and various decorations. What is the most noticeable is the use of saggars in various constellations. Saggar is a lidless clay container, a cylinder, used to protect the glazing of the pottery from flames, dust, and soot during the burning. As soon as the saggars, after some time of use, break, they have to be dismissed from ceramic production. In Chenlu, they are recycled as construction material. Due to this practice, Chenlu has a distinctive characteristic and aesthetical expression, appreciated by the local population as well as visitors and tourists. Today, recirculation of materials is a crucial element in the discussion of environmental problems. In the case of Chenlu, other issues around the utilization of waste materials should also be mentioned: The re-used ceramic elements represent the traditions of the village, reminding the inhabitants of their family histories and the skills passed down from their ancestors, thus shaping and strengthening the identity of the village. Surrounded by the brick and ceramic "landscape," one can better grasp the history of Chenlu.

The houses of Chenlu are situated in a cavity in the landscape, surrounded by farmland. This is a part of the administrative unit of the commune, as well. In the calm silence of the village, sounds are easily echoed and carried around. Sometimes the sad and shrill music from burial ceremonies dominates; other days one can enjoy the happy sounds of children singing floating out through the open windows of the school. Both types of sounds underline the fact that the parent generation has more or less left children and grandparents behind in the village.

The school is a boarding school with children from other villages living in dormitories. A young teacher we met told us with tears in her eyes that many of these children were too small to live separated from their relatives. "They need more care than we are able to give them," she said. To ensure she had a job when she finished her education, she applied for this job. Now she felt abandoned, without contact with her friends and old classmates who worked in cities. The schoolmaster wanted her to stay. She was a sad, conscientious and lonely person. We did not see her on a later revisit to Chenlu. Hopefully, she now lives happily somewhere else.

Tourism is seen as a future possibility for economic growth in the village, and several improvements in sanitary and building standards have been made to attract visitors. When visiting Chenlu in 2017 we could observe improvements in the service street and surrounding buildings, a couple of new public toilets in the village side streets, a couple of new restaurants, a new temple for the pottery gods, and signboards helping visitors to find the few family kilns that were still in function.

14

资源友好型
建筑
西安 / 侯冀
1993/
2002/
2014—
2016

14

RESOURCE-
FRIENDLY
ARCHI-
TECTURE
Xi'an/Houji
1993/
2002/
2014-
2016

所谓"生态设计"涉及各种策略和众多要素。我们应如何管理当下资源和未来的资源使用？本章我介绍三个典型案例，以展示解决这一问题的各种方法。第一个案例是在未开发用地上的设计项目，一个潜在的资源友好型项目。第二个案例是对保护方法的考察，需要把资金以外的其他因素均纳入考虑。第三个案例是在提高能源效率的框架下，处理有人居住的历史建筑。除了物理环境本身，社会文化因素也是发展的重要资源。

西安城墙下的住宅设计
（1993年）

在未开发用地设计新住宅区，并尝试把当地的资源和条件相联系时，会面临与处理使用中的居住区不同的问题和挑战。此处介绍的项目是我们在西安实践所谓"绿色建筑"的首次努力。我们通过项目设计检验了一些生态原则。西安市规划局请我们调查邻近北城墙内一块用地的潜力。比较遗憾的是，因为一位私人投资者买下了这块土地用作他途，这个项目被取消了。虽然如此，我还是在这里略作介绍一下。

环绕西安城墙，在护城河与城墙之间为一长条形地块的公园。我们建议利用这个公园里的植被，对新设计住宅区中的中水进行生物净化。我们的项目尝试在这样一处所在，重新阐释并创造一个现代版本的传统院落系统，有充足的光线和空间。文中图14-2、14-3、14-4、14-5表达了该设计的一些主要原则。

资源流动研究项目
（2002年）

基于我们在传统院落保护工作中积累的经验，我们建立了一个由挪威研究理事会（NFR）资助的研究项目。参与研究的人员来自鼓楼回民区项目办公室、西安建筑科技大学、挪威科技工业研究院（SINTEF）和挪威科技大学（NTNU）。[21]

---

[21] SINTEF report STF22 A03506, April 2003. 鼓楼街区环境与资源分析——开发一种用于评估城市更新战略的工具。

在比较关于历史街区用地性质和建筑体量的可替代方案时，有一个经常被提出的基于经济合算的论点，即什么是最便宜的，什么是最贵的？与经济因素相比，这类项目的其他方面对于决策者而言往往并不那么重要。我们在评估和比较替代方案时，应如何强调价格问题之外的其他价值？与其狭隘地关注短期的支出，哪些长期问题可以并且应该被考虑？

我们此项研究的重点是基于西安鼓楼回民区项目的一个子项目所形成，该项目是关于传统院落的保护。其基本的前提是：中国的城市发展往往导致历史建筑被拆除，以便使用现代技术和新材料建造新的建筑。但是，我们能否找到一个可持续的解决方案，使这些老建筑得到保存？我们选取了三个当地案例，所采用的方法亦可应用于其他地方的城市更新区域。我们制定了一个相当全面和详细的评估模型。它由三组主要标准组成：环境标准、住房标准和文化遗产标准。所有标准的权重都在 -1~6 的范围内。在 2002 年 5 月的一次研讨会上，我们对该模型进行了讨论。讨论小组成员来自不同群体，如省市有关部门的官员、相关专业领域的专家，以及鼓楼片区的居民。该模型经过修改后付诸实践，用于对保护原有建筑和建设新建筑的情况进行比较。

由于很多指标只能估算，我们当然必须对其结果有所保留。不过，我们的结论是，根据选定的标准和价值，有明确迹象表明，保护和修复现存建筑，是一种比拆除建造同等规模的新建筑更适合城市更新的解决之道。

侯冀村研究项目
（2014—2016年）

在侯冀村项目中，我们同样采用了一种整体介入的方法。*22 从能源效率角度来看，该方法适用于对仍有人居住的具有重要历史价值和建筑价值的老院子进行可能的翻新。如此前所述，传统院落是中

---

*22　译者注：关于平遥侯冀村项目研究方法的介绍，参见：Harald Høyem，陈剑芬译，《价值与方法——以平遥侯冀村研究项目为例》[J].《住区》，总第95&96期，2020 (01,02):30-36。

国文化遗产的重要组成部分。现代大都市快速而剧烈的城市更新使得大量老建筑几近被完全淘汰。而在乡村，则还有不少古老的传统院落佳例。有些保存尚好，有些则相当破败。年轻人多离开村子去他处务工或上学，很多院落空置，或仍有些留守的老人，而老人既无精力也无财力去维护这些珍贵有价值的老建筑。如何才能改善这种状况？即是我们要面对的问题。

侯冀村是山西平遥县众多村落中的一个，这些村落在中国历史上的经济贸易活动中都曾留下重要印记。平遥在清代是全国最有影响力的金融中心。很多银行票号富商在平遥周边建起村庄，部分为其家人、员工、仆从居住，部分为耕作务农所用。今天侯冀村位于平遥新高铁站附近。住在这里的村民们都对历史名城平遥的发展和火车站周边区域的规划充满期待。

平遥当地政府邀请东南大学就其古院落在未来的修复和发展可能性进行研究。东南大学与挪威科技大学合作，后者从挪威研究理事会（NFR）获得了一笔能源效率项目的资金支持。能源效率利用的研究框架进而被扩展，把社会文化资源纳入其中。该研究分为三个工作组，每个小组均由来自东南大学和挪威科技大学的研究人员组成。第一组负责居住遗产建筑的保护更新；第二组负责调研了解当地生活和相关知识信息，作为保护和现代化更新工作的前提；第三组负责气候条件分析和环境效能的研究。

调查显示，除砖墙从地下吸水受潮被侵蚀，一些屋顶为植被破坏，项目涉及院落的基本状况良好。通常情况下，每年清明节（4月4日、5日）家中亲戚回村里扫墓时，都会帮助打扫修整院落屋面。由于这一传统正在逐渐消失，杂草丛生，占据了屋顶，而留守村中的老人们也无力处理此类问题。

我们团队人类学家的报告显示，根据传统，男方父母会在儿子结婚时为其准备一套婚房，是为一种荣耀。这样做的结果是，父母一代留在老房居住，而年轻夫妇则搬进新建的房子。老人因财力和身体原因，往往无力维修老宅。年轻人似乎更喜欢住现代化的新房，老一辈则更乐意住老院子。但也有居民表示，半空的院子作为居所让人感到沮丧压抑，尤其是部分建筑因缺乏维护而迅速颓败。对于一个人在院中独居，大家感受不一。在许多方面觉得安全，但也感觉不够安全，因为住在院子里没有和他人的直接接触，更容易发生入室盗窃。想要缓解村中这一问题，扭转衰败趋势，更新、修缮乃至重新安置似乎是一条解决之道。

根据第一、第二工作小组的调研成果，第三组对当地气候和建筑的环境适应性能进行了分析。对房间结构和建筑部件的研究，为计算房屋能量流动提供了数据。通过与居民的交谈，我们了解到院落一年四季的日常使用情况，以及居民如何解决房屋内部的热舒适度。最后，我们在室内外安装了测量仪器和传感器，以观察比较一年四季中室内外的气候波动，包括空气温度、湿度、辐射、空气流动和空气质量。测量结果由东南大学收集并做数据分析。分析呈现出一系列问题：应如何提高房屋的舒适度？应如何在尊重建筑的历史和建筑价值前提下，改善建筑的热工性能？

根据研究结果，我们就如何实施示范项目提出了一系列详细建议。一个至关重要的问题是如何在使用老院落的同时给住户带来收入。不解决这个问题，很难想象如何能保护好传统院落这一美丽而重要的文化遗产。我们就此提出了一些想法，如把房屋出租给在附近村镇工作的人，为游客提供住宿和本地特色膳食，出租给艺术家做工作室、会议研讨室，农家乐（相关家庭在假期和主人家一起体验农家生活），暑期夏令营，微型乡村历史博物馆，等等。实现这些想法需要资金支持、组织细化步骤以及启动相应的试点项目。希望这些策略构想可以帮助老主人、新主人留在他们的院子里，并有条件维护和保护这一宝贵文化遗产。

图 .14-1
平遥县侯冀村
（2015年）

Figure.14-1
Houji Village,
Pingyao
(2015)

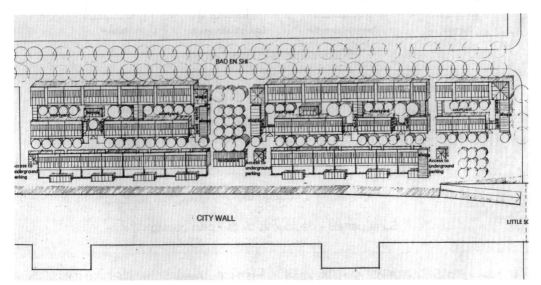

图 .14-2
场地设计

Figure.14-2
Site plan

（图源：西安大宝吉巷片区项
目挪威团队方案，1993年）

(Photo credit: Project report
-Da Bao Ji Xiang quarter,
Xi'an. NTNU, 1993)

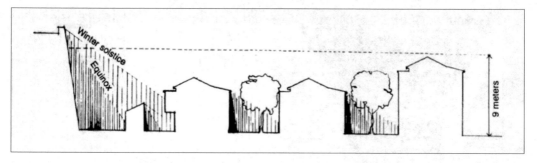

图 .14-3
日照分析
（图源：西安大宝吉巷片区项目挪威团队方案，1993年）

Figure.14-3
Solar studies
(Photo credit: Project report -Da Bao Ji Xiang quarter, Xi'an. NTNU, 1993)

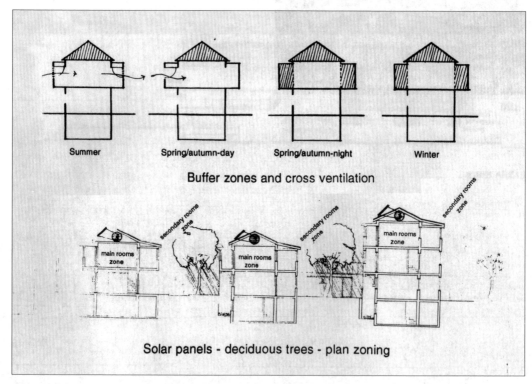

图 .14-4
缓冲区、交叉通风、太阳能板利用、落叶乔木

种植及功能分区分析
（图源：西安大宝吉巷片区项目挪威团队方案，1993年）

Figure.14-4
Buffer zones, cross-ventilation, solar panels,

deciduous trees and plan zoning.
(Photo credit: Project report -Da Bao Ji Xiang quarter, Xi'an. NTNU, 1993)

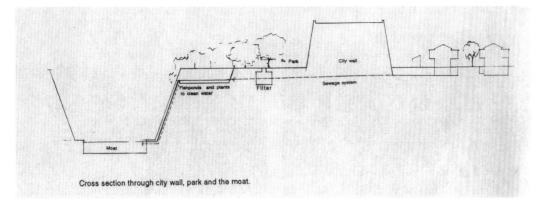

Cross section through city wall, park and the moat.

图 .14-5
剖面研究

（图源：西安大宝吉巷片区项目挪威团队方案，1993年）

Figure.14-5
Sectional study

(Photo credit: Project report -Da Bao Ji Xiang quarter, Xi'an. NTNU, 1993)

图 .14-6
城墙地段典型场景（城墙在画面右侧的树丛后面）（西安，2019年）

Figure.14-6
The typical situation. The City Wall is behind the trees to the right
Xi'an (2019)

图 .14-7
典型院落
（平遥侯冀村，
2014年）

Figure.14-7
Typical
courtyard
Houji, Pingyao
(2014)

图 .14-8
院落—后方远处
为高铁线
（平遥侯冀村，
2015年）

Figure.14-8
Courtyard
roofs –
highspeed
train at the
rear
Houji, Pingyao
(2015)

图 .14-10
破败的屋面
（平遥侯冀村，
2014年）

Figure.14-10
Roofs in decay
Houji, Pingyao
(2014)

图 .14-11
被侵蚀的墙体
（平遥侯冀村，
2014年）

Figure.14-11
Eroded wall
Houji, Pingyao
(2014)

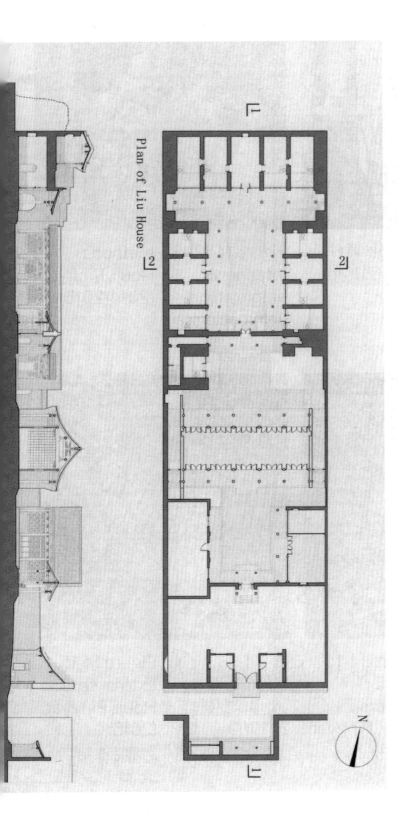

Plan of Liu House

图 .14-9
选作案例研究
的院落

（图源："舒适性之外——侯冀
村传统住宅能源改造"挪威团
队工作报告，2017年）

Figure.14-9
The selected
courtyard for
case studies

(Photo credit: Beyond
Comfort. Energy
retrofitting of an historic
house in Houji Village.
NTNU. 2017.)

图 .14-12/13
以社会人类学
方法进行访谈

（平遥侯冀村，
2014/2015年）

Figure.14-12/13
Interviews
employing
anthropological

methods
Houji, Pingyao
(2014/2015)

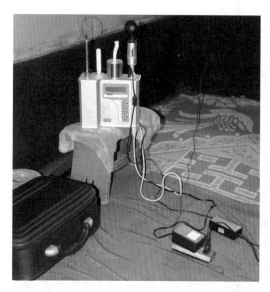 

图 .14-14
室内传感器
（平遥侯冀村，
2014年）

Figure.14-14
Interior
sensors
Houji, Pingyao
(2014)

图 .14-15
室外传感器
（平遥侯冀村，
2015年）

用于测量气候数据并通过互联
网持续传送给东南大学

Figure.14-15
Exterior sensors
Houji, Pingyao
(2015)

Equipment for measuring
climate data, which
continuously were transferred
to SEU by internet

图 .14-16
沉睡的古院
（平遥侯冀村,
2015年）

Figure.14-16
A sleeping
beauty
Houji, Pingyao
(2015)

图 .14-17
屋顶景观
（平遥侯冀村,
2016年）
村中许多富于历
史价值和建筑品
质的传统老院落
都处于类似自然
劣化威胁之中

Figure.14-17
A roof
landscape,
Houji, Pingyao
(2016)-A
sentimental
picture,
maybe, but
illustrates

that many
traditional,
old courtyards
of high
historical and
architectural
qualities are
threatened by
decay

So-called "Ecologic design" involves various strategies and numerous elements. How do we administer the present resources and the future use of resources? In this chapter, I present three typical cases, showing various approaches to the problem. The first one is a design project on undeveloped land, where the circumstances allow for a potentially resource-friendly project. The second is an investigation into conservation methods when other aspects than the finances are taken into consideration in. The third project concerns the handling of currently inhabited and historically interesting buildings, in the framework of energy efficiency. Here the socio-cultural factors – in addition to the physical environment itself – is an important resource for development.

A HOUSING DESIGN PROJECT
LOCATED INSIDE THE CITY WALL OF XI'AN
1993

When designing new housing areas on an unbuilt site, while trying to relate to the local conditions and resources, there are other challenges than there are when working in an inhabited area. The project described here is our first effort to practice so-called "green architecture" in Xi'an. Through project design, we tested some ecological principles. The City Planning Office asked us to investigate the potential of a site inside and adjacent to the northern City Wall. Unfortunately, the project was canceled because a private investor bought the land for other purposes. This was a pity – this could have been an interesting experiment. Although the project was canceled, I'll give a description of it here.

Around the City Wall of Xi'an, there is a thin strip of a park zone between the walls and a moat. We proposed using the vegetation in this park as a biological purification of grey-

water from a new housing area. Our project tried to reinter-
pret and create a modern version of the classical courtyard
systems in a location that, to us, seemed to have decent
amounts of light and space. The sketch below shows some
principles of the design.

THE RESOURCE FLOW
RESEARCH PROJECT
2002

Building on the experience we had accumulated through
the work on the traditional courtyards, we established a research project financed by the Norwegian Research Council, (NFR). Participating researchers were staffed by the DT-
MD project office, the XAUAT, the Foundation for Scientific
and Industrial Research (SINTEF), and the NTNU. * 11

There is an economic argument to be made (and it is fre-
quently made, too) when comparing alternative use of sites
and building volumes in historic districts. What is the cheapest, and what is the most expensive? Other aspects of such
projects tend to have less weight for decision-makers. How
can we emphasize other values than price when we evaluate
and compare alternatives? Rather than a narrow focus on
the short-term price tag, which long-term issues could and
should be taken into consideration?

The research focus was shaped by one of the DTMD
sub-projects, one which looked at the conservation of traditional courtyards. The basic premise was: Urban develop-

---

* 11   SINTEF report STF22 A03506, April 2003. Environmental and Re-
       source Analysis of the Drum Tower District. Developing a Tool for the
       Assessment of Alternative Urban Renewal Strategies.

ment in China often leads to the demolition of historic buildings in order to construct new ones with modern technology and new materials. But can we find a more sustainable solution, enabling the preservation of these old buildings? Three local cases were selected, and the method utilized could be applied in urban renewal areas elsewhere. A rather comprehensive and detailed evaluation model was elaborated. It consists of three main groups of criteria: Environmental criteria, housing standard criteria, and cultural heritage criteria. All the criteria were weighted on a scale of minus 1 to plus 6. The model was discussed in a workshop in May 2002. The members of the panel discussing the model were drawn from different groups, such as politicians and the relevant authorities at provincial and municipal levels, experts in relevant professional fields, and residents from the Drum Tower district. The model was modified and then put into practice by comparing conservation and rebuilding modern houses.

Because there are many indicators that can only be estimated, we must of course express some reservations about the results. We did, however, conclude that there were clear indications, according to the selected criteria and values, that the preservation and restoration of existing buildings is a more suitable solution for urban renewal than demolishing and constructing new buildings of comparable size.

THE HOUJI VILLAGE
RESEARCH PROJECT
2014-2016

In the Houji Village project, we also employed a holistic approach. From the perspective of energy efficiency, this applies to the possible refurbishment of old buildings in inhabited courtyards of great historical and architectural value. As mentioned earlier, original, traditional courtyard compounds are an important part of the cultural heritage of China. The fast and dramatic urban renewal of big cities and metropolises has caused an almost complete elimination of the old buildings. In the countryside villages, there are, however, many splendid examples of old, traditional courtyards. Some are in relatively good shape, while some are quite dilapidated. Young people have left to find jobs and schools elsewhere, and the courtyards are now empty or inhabited by old relatives who do not have the energy or financial means to maintain the valuable buildings. What can be done to improve this situation? This was the problem we wanted to confront.

Houji is one of many villages in Pingyao County, villages with visible footprints of the famed economic history of China. During the Qing Dynasty (1644-1912), Pingyao was the most influential, financial hub of the country. Rich bankers built villages, partly for their own families, partly for their staff and servants, and partly for farming, in the landscape surrounding Pingyao city. Houji is located near the new high-speed train station of Pingyao. The villagers are waiting for the development of Pingyao historical city and the planning of the district surrounding the railway station.

The Pingyao government asked SEU in Nanjing to carry out research on the potential of the classical courtyards for future restoration and development. And SEU joined forces with the NTNU. NTNU received funds from the Research Council of Norway (NFR–a state unit), under an energy efficiency program. The framework of energy efficiency was extended to include socio-cultural resources. The research was organized into three groups, each manned by researchers both from SEU and NTNU. Group 1 dealt with conservation and refurbishment of heritage dwellings; group 2 with local life and local knowledge, as a premise for conservation and modernization; group 3 with climate analysis and environmental performance.

Investigation showed that the courtyard in question was in good shape, except that brick walls eroded because they sucked water from the ground, and roofs were destroyed by vegetation. Normally roofs would be repaired by relatives when they came to the village during the "Sweeping the tombs festival" (4th/5th of April). They then would look after the family tombs and clean the roofs of the family courtyard. As this tradition is vanishing, vegetation freely attacks the roofs, because it is too hard for the older residents who remain in the village to manage the problem by themselves.

Our team's anthropologist reported that according to tradition, it has been honorable for the parents to provide their son with a house when he marries. A result of this is that the parents' generation stays on in the old house and the young couple move into a newly-built house. It is difficult for the old generation to keep up with maintenance, both because of the financial burden and reduced physical strength. It also appeared that young people preferred to live in a new, modern house, whilst the older generation was more satisfied with the old courtyards. However, some residents described the half-empty courtyards as depressing places to live, especially when parts of the building were quickly deteriorating because of lacking maintenance. There were mixed feelings about living alone in a courtyard. In many ways, it feels safe, but it also feels unsafe, because from the yard there is no direct contact with other people, and so residents would, for instance, be more vulnerable to intruding burglars. Upgrades, repairs, and repopulation seem to be a way to reverse the trends and alleviate the village's problems.

Based on the findings of Groups 1 and 2, Group 3 worked with climate analysis and the environmental performance of the buildings. Studies of room structures and building components gave data for calculating the energy flow. Dialogues with residents gave us an understanding of the daily use of the courtyard throughout the year, and how residents obtained thermal comfort. And, finally, measure instruments, sensors, were installed indoors and outdoors to observe and compare the fluctuation of the climatic factors inside and outside – air temperature, humidity, radiation, air movements, and air quality, during all seasons of the year. The measurements were transferred to Nanjing, where SEU collected and analyzed the data. From the analysis some questions formed: How can the comfort in the houses be improved? How can the thermal performance of the buildings be improved, while still respecting the historical and architectural values of the structures?

Based on findings from our research, we offered a series of detailed recommendations for carrying out demonstration projects. The crucial, superior problem will be to find ways to use the old buildings that will also help generate income for the owners. Without solving this problem, it is hard to imagine how it will be possible to protect the beautiful and important cultural heritage of the traditional courtyards of China. We came up with some ideas, like renting out rooms for people who are working in villages and towns nearby, accommodation for tourists including meals with local food, restaurants also with local food, ateliers for artists, seminar rooms, farm tourism (i.e., families staying with the hosting family working together with them during the vacation), summer-school, small museum for local history, and so on. Ideas will have to be substantiated and elaborated; organization and financial structures have to be established, and test projects initiated. With such a strategy, it would hopefully be possible for original or new owners to stay in their courtyards and finance the maintenance of this valuable cultural heritage.

15

重新审视
中国
2018—
2019

15

CHINA
REVISITED
2018-
2019

动手写此书时面临一个两难问题，即谁是读者。对于一些非中国读者，或是希望笔者能为其提供有关中国的新知识。而对有些中国读者，许多内容于他们不过是司空见惯的常识，但他们亦有兴趣了解一个外国人所关注和观察到的中国。我试图同时满足这两类读者的要求，但必须承认，这是一个棘手的平衡。本章的重点是六个主题：城市化、交通、污染、食品、商贸、文化。它们代表了我在1985—2019年期间所经历的大部分事情。

自1985年我首次到中国以来，中国发生了哪些变化？又有哪些东西似乎没有变化，甚或永久不变？我将在此简要总结一下自己对某些问题的观察和思考。中国在过去30多年里经历了巨大的变化，其发展的规模和速度让人印象深刻，甚至可以说是令人震惊。这种超高速的发展是如何实现的？在这一过程中得到了什么，失去了什么？下面的文字无意对这些问题给出完整的答案，而是尝试通过我在中国参与的事和简单的观察，描述我的亲历体验，以及我所看到的不变和变化之处。

首先是一个所谓的"厚重描述"（Thick Description，社会学术语，指对于复杂文化现象的描述性诠释），起始于物质环境的变化。其推动力是人们从农村向城市的迁移，可以称之为移民时代。1978年以前，中国的城市化速度比其他第三世界国家要慢很多。但随着后来经济改革的升级，以及新立法允许人们更加自由地流动，数以千万计的农村人口进入城市，并定居于此。农村缺乏有偿工作机会、农产品价格低廉，与城市相比，教育和医疗服务落后。——所有这些因素，首先吸引了个别农民工来到快速发展的城市，然后，这些人最终把整个家庭迁徙到城市。中国的城市化以惊人的速度发展，城市人口从1982年的20%增长到今天的50%左右。北京的情况可以形象地说明这一点。北京人口从2007年的750万增长到2015年的2100万。这一变化部分是由于城市行政单位边界的调整，尽管如此，北京的人口增长还是极其惊人的。

人们从乡村数以百万计地涌入城市，因为没有城市户口，他们在很长一段时间内无法享受到城市常住居民所享有的公民权利。不断增长的人口让城市面临着巨大的压力，与此同时，农村人口的减少削弱了小城镇和乡村的生存能力。

就我个人观察，变化是逐步发生的，也是迅速发生的。我曾经问过一位在北京从事住房开发工作的高级官员，在很短时间内，人们从胡同搬到高层住宅新区，曾经生活在胡同文化氛围中的居民是否会有困扰。他回答说，是的，他们很清楚一些问题的确存在，但必须容忍这些问题。他们的主要任务是为城市中生活在恶劣居住条件下的人群和从农村迁入的人群建造住宅，因此其他的困难与问题只能先搁置到一边。我们很难知道搬迁对百姓的影响是什么。可能经过几代人的时间，会更容易看到这些年发生的变化所带来的结果。

城市化

让我们先粗略勾勒一下过去25年里发生了巨大变化的城市景观面貌。最显著之处是城市规模的增长：地块的使用、城市街区的大小，以及建筑物的高度。面对不断扩张的环境，只有像北京的故宫和天坛那样的历史遗迹才有足够的气势保持其在城市景观中的重要性。

1985年，没有广告灯箱和路灯，夜间在外面活动相当危险。特别是下水井，因为井盖经常被人偷去当废铁卖，在黑暗中骑自行车时有可能掉进下水井。那时候没有私家车，大家都坐公交车或骑自行车。今天这一切都改变了。灯光师、设计师们弄出梦幻般的广告，这些广告可以变换色彩、闪光、切换内容。总之，他们利用一切能想到的高科技手段为市场营销创造幻象。十字路口都设有红绿灯，街面上车水马龙。私家车种类繁多，不乏高档豪车。现在中国人可以购买到一切，包括奢侈品，一些商品的价格比欧洲还要昂贵。手机、电脑、电视和其他电子设备的使用比世界其他地方更广泛普及。如同第一批可口可乐在20世纪80年代在中国上市销售时新闻所说的那样"他们终于跟上潮流了"（虽然那个时候中国跟随潮流的程度还非常有限）。

20世纪50年代以"先生产后生活"为口号的政治纲领，为后来的住房设定了某种程度上的低标准。家庭通过工作单位分得一套公有住

278

房，并支付低廉的象征性租金。通常情况下，住宅都是在工厂、办公室、大学或其他工作地点附近。

在高速城市化进程中，大部分历史街区被拆除，代之以现代建筑。根据20世纪90年代的政治目标 "2000年进入小康社会"，有经济能力的人寻求面积更大、卫生和基础设施标准更高的住宅。房地产逐渐主导了新区的建设。政府批准了大量的开发区，面对快速城市化和改善住房条件的政策所带来的不断增长的住房需求，房地产开发迅速发展。投资建房获利颇丰，新住宅的标准大幅提高，然而小康家庭和低收入家庭之间存在巨大的差距。

过去40年来惊人的发展速度既有积极的影响，也有消极的影响。一方面，其速度足以让人们见证和体验到真正的变化，为人们日常生活带来非常重要的改善；另一方面，却没有足够的时间来反思这种发展带来的诸多影响。也许，经济增长放缓并非坏事？这样，人们会有更多时间来思考和预判当下发展可能产生的影响。

我们可以把城市形态理解为新移民时代的部分结果。一波又一波的人潮不断从农村迁往全国各地的城市中心和大都市，给城市土地和基础设施带来巨大的压力，同时也使得农村的人口、就业机会和社会资产越来越少。在城市中，通常25~30层的高层住宅楼群取代了4~6层高的多层和低层院落小区，新的物理环境在这一过程中的引入也影响到过去的社交网络模式。居住面积增加了，卫生条件更好了，居住标准得到了提升。写字楼如雨后春笋般出现，其中一些是极高的摩天楼，一些由国际公司设计：先是像森林中的蘑菇，一栋接一栋地出现在低层建筑为主的城市组织肌理中。后来，一整个街区的高层建筑完全取代了原有的城市街区，或是在城市郊区的农田上建起了整片的新建筑。在这摩天楼时代的最初几年，建筑的规模和高度已令人瞠目。在建筑奇想和房地产开发驱动下，各种新奇特异的建筑形式应运而生，以吸引投资者。

中央政府负责对各个城市的长期总体规划进行深入审查，特别是在土地使用方面。如果中央认为土地使用不当，报批的规划就会被退回。高密度的高楼大厦经常受到西方人的批评，认为这种发展不够人性化。这种批评或有一定的道理，但我们也必须记住，还有一些非常强大的力量在发挥作用。新居民涌入城市，对城市土地造成压力。将农田用于城市用途意味着可耕种土地的减少。在20世纪90年代，我们经常在报纸上看到由于城市化和沙漠化而丧失的农田数量。通过计算农业土地的减少，就可以计算出人均可获得的粮食量。念及我自己的国家对减少用于粮食生产可耕种土地不甚负责任的态度和做法，中国的这种公共信息透明度让我艳羡。当时中国的政策是努力争取粮食自给自足。后来，当中国增加出口工业时，这一政策发生了改变，但主旨仍是尽可能少地使用土地进行城市发展。当然，政策实施中也有很多例外情况，但至少它为在城市中大规模建造高层建筑提供了一个论据。如果从人口趋势来看，这样做似乎是合理的。另一种选择，是通过分散居住区减缓城市发展压力，而这在目前来看还很难实现。为理解一切何以如此，我们需要从城市化和粮食供应角度来看待这种现代居住方式的发展。当然，当我们使用理解这个词时，并不一定意味着接受。

我们在非洲和拉丁美洲城市中所见到的贫民区，在20世纪八九十年代的中国是不存在的。透过欧洲人的有色眼镜观之，城市中最古老、人口稠密的街区通常是贫民区，且往往被定性为贫民区。的确，中国这些街区的住房质量很差，但这里有自来水，有公共厕所和浴室，有污水处理系统，有垃圾处理，附近有学校和卫生站——或许有些地方看起来很破旧，但至少有基础设施。如果继续深入了解，会对人们在简陋条件下仍能一定程度地体面生活留下深刻印象（当然，这也是全世界贫民区的一个显著特征）。要知道，这并不意味着一切都是田园诗意般的和谐。1990年，我们采访居住在西安市中心鼓楼回民区的居民，那里的人均居住面积是2.7平方米，利益冲突和争吵司空见惯。典型的冲突，是有人想从住宅之间的公共空间中抢占上几平方米（参见第四章）。空间稀缺造成的结果是，人们需要以不同于我们这些来自人口稀少、空间资源充盈环境者所习惯的方式相处。

我们所研究的农村地区的村庄，也已经发生了各种不同变化。有的破旧不堪，处于半废置状态；有的则情况良好；有的处于发展失衡状态，让人怀疑是否应继续存在。另有一些则双管齐下，一边维持着传统营生，一边发展新兴旅游产业。

交通运输

2018年，我为一小群友人规划行程，我们打算乘火车环游中国。通过互联网，我们在中国铁路网站上订购了车票。由于旅行，我们要在很多站中转休息，也因为同行玩伴有不同的愿望和时间安排，因此行程订单相当复杂。所有这些都由接单者妥善处理，并顺利安排了付款。在我们抵达中国前三天，我们订购的车票就送到了酒店前台（那是一个几乎没有专业前台的小酒店），所有的车票都井然有序。我们的第一段旅程是从北京到上海，坐的是高铁。这与我第一次来中国旅行时的蒸汽机车硬座相比，真是日新月异的变化。

在改善铁路系统的同时，其他长途运输交通基础设施也都得到了扩建，高速公路、桥梁，还有航空运输设施。20世纪90年代，中国航空公司数量很少。中国民航的缩写"CAAC"有时被游客戏称为"航班总是取消航空公司"（China Airways Always Cancelled），这也许有点不太公平，但也与事实相差不大。现在中国民航有多家航空公司，拥有国际化高标准的飞机，运营着无数国内、国际航线。新的机场和航站楼不断涌现，通常由国外知名建筑事务所操刀设计。

新的城市发展带来了新的交通问题，需要新的交通系统。住房和工作单位原本建造在工作企业附近，由企业负责为职工提供住房。改革开放政策实施后，住房商品化，人们的居住地开始远离工作单位。由此产生了对机动交通的需求。城市规模的扩大也需要新的交通解决方案。在20世纪90年代的西安，骑自行车出行是主要交通方式。十年后，由于汽车交通繁忙，街道和街道旁边的停车位，甚至是人行道上都停放着私家车，骑自行车变得很危险。20世纪90年代初的骡马运输很快就被货车所取代，噪声和灰尘在街道的空气中留下了痕迹。为了解决这些问题，提高城市交通效率，政府建造了功能完善的地铁系统。与世界上许多大城市一样，中国大都市的地下交通系统的客容量有限，但新的地铁线路在不断增加。在历史古城中，建造地铁过程中经常会在地下10~20米有考古发现，如何在不影响地下文物的情况下修建隧道是一个难题。以西安为例，地下的考古发现一定程度上延缓了地铁线路的建设。

现在出租车行业基本上使用的都是普通小汽车，但三轮摩托和人力出租车在个别地方仍然存在。在20世纪90年代，出租车数量众多，乘坐出租车出行非常方便。而出租车司机竞相拉客，有时因为争抢客源采用不当手段造成事故。那时为了省钱，大多数中国人首选公交车。近年来出租车行业已然发展成为卖方市场。在一天中的大部分时间，人们必须非常耐心或运气够好才能打到一辆出租车。虽然越来越多的人有能力坐出租车，公交车仍然是最普遍的城市公共交通工具，公交车站常常人满为患。渐渐地，排队等车的文化已经发生了变化。早先人们必须手脚并用争抢着挤上一辆公交车，而现在则秩序井然地排队，不会再让最粗鲁无礼、势大力沉者捷足先登。

污染

在中国很多地区，因为北部沙漠和平原地区土壤侵蚀产生的尘土被吹散到全国各地，空气污染一直是一个大问题。这种状况已经持续了数百万年，也许直到土地不再沙化而成为绿色的时候才会结束，我相信这绝非朝夕之事。多年以来，尘土堆积沉降，有些地方形成的黄土层深达200~300米。

除了来自空气输送的污染，私家车数量的不断增长对空气质量也影响甚大。车辆排放的尾气和交通扬尘，加之工业污染的烟尘，给人的健康带来一系列问题，令政府决策者十分头痛。20世纪90年代初，也出现过尘土飞扬的空气，但如上所述，就像几百万年来的状况一样，这主要是源于北方沙漠和平原吹来的颗粒物。随着人为空气污染的加剧，情况已经恶化到了难以忍受的地步。即使只是短暂旅行逗留，浓重的雾霾也是让人深感压抑。污染的日子和晴空万里下城市风景一览无余的时日相比，有着云泥之别。很多西方媒体认为中国政府并不关心这一问题，这当然并非事实。实际上，和世界任何地方一样，环境污染是一个巨大且难以解决的问题。要解决这个问题，一项实质性的方案就是减少城市中私家车的使用。中国政府已采取了一系列或严厉或温和的举措。这些措施确实收到了一些效果，因为普通公众现在也明白这种环境危机，正如很多人亲身体会到的呼吸系统问题。人们或许也意识到，在交通堵塞中卡上数小时，距离汽车企业广告中所许诺的行动自由相距甚远。

欧洲城市也面临着同样的问题，只不过程度没有那么明显和普遍。因此，从国外乘航班进入中国，当飞机一头扎入城市上空的雾霾中时，这种反差总是让人压抑。中国政府通过不同的举措努力改善空气质量，如在交通系统、能源使用供应、关闭污染工厂、家用能源使用现代化，等等。近几年，我也亲身感受到空气质量的改善：最近两次去北京期间，天空一直很晴朗，可以看到城市北边的山峦。统计数据也显示，尽管问题依然严重，但在向好的方向发展。在全球应对气候危机的斗争中，这些都是令人鼓舞的迹象。

食品

在我访问中国的头几年，食品都是定量供应的。就像二战后的挪威那样，每个家庭都会配发食品券。那时中国家庭用粮票购买基本日用食品，商品的价格都在普通家庭可承受范围。外国游客则没有这个问题。餐馆的食物很便宜，按中国的饮食文化方式烹饪。但普通人没有能力每天下馆子吃饭，家庭主妇、男人们要在菜市场花时间寻找最便宜的食材。那时仍然有非常穷困的人在温饱线上挣扎，但大多数城市居民可以在商场买到日常所需，而且相对家庭开支商品价格也比较合理。农村地区的农民经济条件则差一些。我有次采访一个农民家庭，他们说缺少钱款维持日常开支，但因为自己种地，所以尚能饱腹。我没有到过偏远地区耕作困难且极度贫困的乡村，因为没有第一手的亲历观察，我当然也不能说不存在严重的缺粮少食问题。

有时，好客的人们会邀请我到家里吃饭。看他们如何做出健康美味可口的饭菜是一种享受，吃饭气氛总是友好而融洽的。外国客人经常会被邀请到餐馆和几个朋友一起聚餐，或是更正式的宴会场合。我注意到这类饭局的一个变化。早些年，无论饭菜质量还是菜品数量都让人应接不暇。近几年来，饭菜标准发生了改变（我本人热烈欢迎这种变化），饭菜变得更简单——提供的菜品少了，肉类少了，蔬菜多了。其原因是政府为打击腐败，出台了一些规定，限制公务餐的数量和花费。

商贸

当人们无法自给自足的时候，商品的买卖交换就成了必需。在过去的30~40年里，商品买卖的方式确实发生了变化。在我初次来中国的时候，外国人会收到特殊的货币——外汇兑换券（FEC），允许持有者在友谊商店中购物。在这些商店，人们可以买到普通商店里没有的东西。外汇券在当地中国人中也非常受欢迎，因为普通商店中的商品种类极其有限。很多人有钱也无处可花。今天中国的城市就像世界各地的城市一样。大城市中的商店物品种类琳琅满目，城镇和乡村的店铺则有日用必需品。店面广告充斥街道，改变了城市景观的视觉意象。只要有钱，什么东西都可以买到。

20世纪90年代初，西安只有中国银行为外国人提供服务。外国人必须在那里把旅行支票兑换成外汇券。如果要进行重要的转账，就必须去北京的中国银行总部。现今，城市里到处是银行大楼，它们通常都是城市里最气派的建筑，而且每个街角都有自动取款机。

网络购物正不断升级。在大学校园里，学生们在手机上点击一下，就可以订购任何东西。几天后，包裹被装上电动三轮车送到校园门口，摆放在人行道上，学生们过来取走包裹。当场试穿衣服，如果不满意，包裹可以由送包裹的人退回。完全是一种高科技数字交易与低技术人力配送的结合。

近几年来，互联网贸易发展起来。越来越多的人在网上购物，这也影响到商业购物基础设施。我的一位同事指着一栋相当新的十层建筑告诉我说，几年前整栋大楼是一个电子产品市场。现在这栋建筑已空空如也，被互联网商业所取代。这位同事自己也喜欢在网上购物。他住在北京五环外，如果要开车去市中心，在路上会堵上好几次，然后还得找停车位。一趟商场之行几近要花费他一整天的时间。他所在的社区有几家不错的餐馆。大多数人都选择在这些餐馆点餐，然后由快递员送餐到家。

文化

文化是一个全面的概念，渗透在生活的各个层面。我仅选择几个我认为可以描述中国社会特点的问题。这样做部分是缘于它们乍看似乎与我自己所在的文化不同，但仔细思量，可能也是我们的共同之处。

在任何国家，文化都被用于政治和商业的目的。文化可以用我们所谓"软实力"的方式被运用。例如，2013年发起的"一带一路"倡议，最初的丝绸之路一直是所有欧洲人从马可·波罗的冒险中知道的故事的主干。2004年正式启动的孔子学院是另一个例子，它充分利用了我们西方人对生活在公元前551年至公元前479年的孔子的模糊认识。孔子学院传播了有关中国的信息，现在已经在全世界范围建立了良好的基础。和所有国家一样，文化为政治所用，是一个国家及其民众不断形成身份认同的一个有效工具。在将身份认同作为社会发展的建设性因素时，似乎很难在民族自豪感和民族沙文主义之间找到一个平衡。西安的两个考古遗址公园——唐大明宫遗址和汉长安城遗址（见第十章）的建设，就是为了展示国家辉煌历史而进行的巨大财政和政治投入的实例。这些考古公园的设立，对于吸引游客到西安旅游，让当地和国内游客形成历史自豪感非常重要。

作为十三朝古都，西安及其邻近地区有许多文物遗址，自20世纪50年代以来有许多考古发掘。1974年意外发现了著名的兵马俑。此后，博物馆和参观遗址的各种方式发展起来，以吸引中外游客。我第一次去参观兵马俑，是1989年，当时兵马俑以简单、朴实无华的博物馆学方式展出。后来，为满足不断升级的旅游业和文化遗产保护专业要求，展陈方式越来越精细。当年我是骑着自行车一路颠簸去参观的。如今游客们则乘坐大巴、出租车在四车道的高速公路上前往。那里又有了更多的考古发掘和新发现。新的博物馆纪念建筑在遗址周围拔地而起，凸显着这·世界遗产的重要性。

其他历史名城的发展也类似。值得注意的是，很多年前国内游客的数量就已经远远超过了国外游客。政府鼓励国内旅游，人们排着长队去那些著名景点参观。比如我在2018年4月去北京的时候，尽管有非常高效的售票检票系统，进入故宫的排队等候时间依然长达一个小时。在旅游旺季，像国庆节前后的假期，往往人满为患——对于很多远道而来、想充分利用假期的人来说，把大把时间花在排队上无疑让人沮丧。这种体验与我1985年初次来中国旅游时截然不同，那时国内游客很少，出现一个国外游客更是引人注目，游客们往往彼此凝视，相互打量。

西安曲江大雁塔周边区域的发展，代表了城市发展的新模式。它利用历史和文化古迹来吸引投资，推动城市生活。大雁塔是几经修复的唐代古迹。1985年，它在城市的南端，周围农田环绕，紧邻着大慈恩寺。当时我骑着自行车，穿过一片田园牧歌式的宁静风景，到佛塔附近时，那里没有游客。如今，这里熙熙攘攘，有餐馆、电影院、博物馆、购物中心、剧院、公园和各种公共设施，包括一条高架轻轨线。这片区域是一种特殊规划方法的成果。西安市政府在启动该片区的开发时，首先投资建设必要的基础设施，在这些条件具备后，又利用历史作为市场营销工具，吸引开发商到这里投资。为了展现历史，大雁塔当然是一个重要的组成内容，此外还有一些作为背景线索的考古遗迹，无数镌刻着古诗词的新石刻，以及描绘重要历史人物的纪念性雕塑。曲江片区是西安最近的一个城市级规模的实验，在这里，历史与商业利益紧密关联。

造访中国之所以如此有趣，原因之一是那种古老传统和现代表达的混融。例如，中文书写系统对于一个外国人有种异域特色和视觉观赏品质，而对中国人来说，它是国族身份认同的重要组成部分。我听说有八万多个汉字，今天聪明的人们已经实现在个人电脑、手机上书写和阅读这些汉字，所有这些汉字在触屏上书写得飞快。斯堪的纳维亚语言有29个字母，英语有26个字母，而斯堪的纳维亚语言中的三个特有字母 æ、ø 和 å 至今仍未完全被引入到数字化系统中。这与中国文化的高速现代化形成巨大的反差。

图 .15-1
城市信息屏
（北京，2019年）

Figure.15-1
Horizontal
information
screen
Beijing (2019)

图 .15-2
高楼林立、
车水马龙的城市
（上海，2019年）

Figure.15-2
A city of high-
rise buildings
and heavy
traffic
Shanghai
(2019)

图 .15-3
拥挤的地铁
（北京，2019年）

Figure.15-3
Inside subway
Beijing (2019)

图.15-4
山西某工业城镇
（山西，2013年）

Figure.15-4
Factory town
in Shanxi
Province
Shanxi (2013)

图.15-5
路边风干肉
（柞水县凤凰镇，
2009年）

Figure.15-5
Drying
meat on the
sidewalk
Fenghuang,
Shaanxi (2009)

图.15-6 侨福芳
草地购物中心
（北京，2019年）

Figure.15-6
Parkview
Green
shopping mall
Beijing (2019)

图.15-7
老年街头艺术家
（北京，1995年）

Figure.15-7
Senior street
art painter
Beijing (1995)

图 .15-8
大雁塔旁的
曲江新区
（西安，2011年）

Figure.15-8
Qujiang
District, near
the Big Wild
Goose Pagoda
Xi'an (2011)

图 .15-9
浦东金茂君悦
大酒店中庭
（上海，2018年）

Figure.15-9
Atrium of
Grand Hyatt
Pudong
Shanghai
(2018)

图 .15-10
道教名胜齐云山
景区游客通道
（黄山，2018年）

Figure.15-10
Access to the
Taoist Qiyun
Mountain
Huangshan
(2018)

图 .15-11
繁忙的北京南站
（北京，2019年）

Figure.15-11
Busy Beijing
South Railway
Station
Beijing (2019)

图 .15-12
作为城市景观的
现代十字路口
（济南，2019年）

Figure.15-12
Modern
intersection
as urban
landscape
Jinan (2019)

图 .15-13
首钢工业遗产园
区，2022年冬
季奥运会及残奥
会组委所在地
（北京，2019年）

Figure.15-13
Planning
center of
Beijing
Organizing
Committee for
2022 Winter
Paralympics,
Reuse of an
industrial
district
Beijing (2019)

图 .15-14
西安鼓楼回坊北院门街小吃店
（西安，2013年）

Figure.15.14
Snack Shop at Beiyuanmen Street, DTMD Xi'an (2013)

图 .15-15
酒店大堂一隅
（西安，2019年）

Figure.15-15
Hotel reception Xi'an (2019)

图 .15-16
购物中心的3D全息投影
（北京，2019年）

Figure.15.16
3D holographic projection in shopping center Beijing (2019)

图 .15-17
电子产品专卖店
（北京，2010年）

Figure.15-17
Electronics
store
Beijing (2010)

图 .15-18
上海浦东新区
（上海，2018年）

Figure.15-18
Pudong New
Area
Shanghai
(2018)

图 .15-19
商务办公楼群
（北京，2019年）

Figure.15-19
Office
buildings
compound
Beijing (2019)

图 .15-20
南京城市景观
（南京，2005年）

Figure.15-20
Urban
landscape
Nanjing
(2005)

图 .15-21
高层住宅
（上海，2005年）

Figure.15-21
High-rise
housing
Shanghai
(2005)

图 .15-22
高铁车站
（北京，2018年）

Figure.15-22
Train station
Beijing (2018)

图 .15-23
首都国际机场
三号航站楼
（北京，2010年）

Figure.15-23
Beijing Capital
International
Airport
Beijing (2010)

图 .15-24
出租共享单车
（北京，2019年）

Figure.15-24
Rental
bicycles
Beijing (2019)

图 .15-25
室外景观雕塑
（西安，2019年）

Figure.15-25
Outdoor
sculptures
Xi'an (2019)

When starting to write this chapter, I faced a dilemma when questioning who is the reader. There will be non-Chinese readers for whom the following chapter hopefully will offer new knowledge. There will also be Chinese readers for whom much of what is in the following text is well-known information, however finding interest in learning what a foreigner has observed and paid attention to. I have tried to comply with both groups, which has been a tricky balance, I must confess. There is a focus on six main issues: urbanization, transport, pollution, food, trade, and culture. They represent much of what I experienced during the period 1985-2019.

What has changed since my first visit in 1985, and since my first longer-period stay in the country? What seems to be unchanged, perhaps even permanent? I will, very briefly, try to sum up some of my observations and reflections on various matters. China has undergone substantial changes during the last three decades, and development of a magnitude and speed that both is impressive and somehow frightening. How has this ultra-rapid development been possible? What was gained and what was lost along the way? The following text has no ambition to give any full answer to these questions, but it rather tries to describe what I personally experienced, by simple observation and through the activities in which I partook in China, and what I see staying the same and what I see has changed.

Here is a so-called "thick description," starting with changes in the physical environment. The driving force has been what could be named age of migration, with people moving from the countryside to the cities. Before 1978, the speed of urbanization in China was much slower than in similar Third World countries. But with economic reforms, which later escalated and new legislation which allowed people to move more freely, millions of people were pulled into the cities, where they settled. Shortage of paid work in the countryside, low prices for farming products, poor quality of education and health services compared to that which was evolving in the cities – all these factors first attracted individual migrant workers to the fast-growing cities, and then, eventually, these individuals moved their whole families along. The urbanization of China took place at an incredibly high speed, causing a growth in the urban population from 20 percent in 1982 to more than 50 percent today. To further illustrate: Beijing, grew from 7.5 million inhabitants in 2007 to 21 million in 2015. This change is to a certain extent due to adjustments to the borders of administrative units, but nonetheless, the population growth in Beijing is overwhelming.

People from rural districts swarmed into the cities in millions. Here they became groups of unregistered job-hunters, people who for a long time did not have the civil rights bestowed on the permanent urban residents. There was intense pressure on the growing cities, and simultaneously there was a depopulation of the countryside, which weakened the small towns and villages' ability to survive.

What I personally have observed is that changes have taken place gradually and rapidly. I once asked a high-rank politician who worked with housing development in Beijing, if there were some human problems for people moving, during a very short period, from the *hutong* area culture to that one of the new high-rise districts. He answered that yes, they were well aware of some problems, but these had to be tolerated. The main task was to build dwellings for people migrating from the countryside and for people living in poor housing conditions in the city. Solving the smaller problems was simply down-prioritized. It's hard to know what the impact has been on the people. After a couple of generations, it will probably be easier to see the long-term consequences of the changes that have happened during these years.

Urbanization

Let us start with a rough outline of the urban landscape which has seen enormous transformations in the last 25 years. Most conspicuous is the growth in scale: in use of the area, the size of city blocks, as well as the height of buildings. Faced with these expanding surroundings, only historical monuments, like the Forbidden City and the Temple of Heaven in Beijing, have enough grandeur to maintain their significance in the city landscape.

In 1985, illuminated advertising and streetlights were non-existent, making it rather risky to move around in the dark. (A particular danger was the manholes, as the lids were often stolen and sold as scrap iron. These holes became deceitful traps when biking in the dark.) There were no private cars – everyone went by bus or by bike. Today, all this has changed. Light technicians and designers play with fantastic advertisements which shift color, flash, turn around and switch content. In short, they create fantasies through marketing by using every thinkable high-tech method. There are traffic lights at the important crossroads, preventing total chaos with all the private cars that are now in circulation. And the cars are often showy and big. You can buy whatever you want if you have money, especially expensive branded goods now produced in China and sold for higher prices than in Europe. The use of mobile phones, computers, TVs, and other electronic equipment is in more widespread use than anywhere else. Acclaimed by the West as a capitalist triumph, the first Coca-Cola bottles went on sale in the 1980s: "Finally they follow." (Though at that time they followed to a very limited degree).

The political programs of the 1950s with the slogan "Production first, livelihood second," had set a somehow low standard for later housing. Families received a publicly owned dwelling through their work unit and paid a low, almost symbolic, rent. Usually, dwellings were located close to the factory, the office, the university, or other places where people worked.

During the high-speed urbanization, most of the historic districts were removed and replaced by modern structures. In accordance with the slogan of the 1990s' political programs "a relatively comfortable life in the year 2000" - those who could afford it looked for bigger dwellings with improved sanitary and infrastructure standards. Real estate gradually dominated the construction in new areas. Huge districts were released by the authorities for this purpose, and large capitalist actors were mobilized to solve the escalating housing needs caused by rapid urbanization and the policy of improved housing conditions. Investment in housing was very profitable, and the standard of the new homes increased considerably, yet with substantial differences between well-off and low-income families.

The speed of the previous 40 years' impressive development comes with positive and negative effects. On the one hand, it has been fast enough for people to witness and experience real changes and improvements that have been important for their everyday life. On the other hand, there has not been time enough to reflect on the many consequences of the development. Maybe a slowing down of the economy could be positive? One would then have more time to reflect and foresee the possible effects of the ongoing development.

One may understand the urban form as partly a consequence of the new age of migration. A constant wave of people who have moved from the countryside to urban centers and big cities all over the country has caused a lot of stress on urban land and infrastructure and has at the same time left rural villages with a dwindling population with a dwindling number of jobs and social assets. In the cities, clusters of high-rise housing skyscrapers, typically 25-30 stories high, have replaced low-rise courtyard districts and areas with 4-6 stories tall houses, and in the process introduced a physical environment that has affected the social network patterns of the past. Dwelling standard has been improved by introducing better sanitary condition and increased living areas. Office buildings, some of them extremely high skyscrapers, some of them designed by international companies, sprouted up; to start with one by one in the low-rise urban tissue, like mushrooms on the forest floor. Later development has completely erased existing city blocks to make place for entire districts of high-rise buildings or has created new constructions on farmland in the outskirts of the city centers. In the first years of the sky-scraper era, their size and height were already sensational. But later, architectural fantasy and the demands of the real estate companies resulted in even more exotic forms to attract the interest of investors.

The long-term master plans for the cities are thoroughly investigated by the central government in Beijing, particularly when it comes to the use of land. The plans will be rejected and returned if the central authorities consider the land use irresponsible. The high-density, high-rise movement is often criticized by Western people who'll describe it as inhuman and cynical. There might be something to this critique, but we also have to keep in mind that there are some very strong forces at work. The flow of new residents to the cities causes stress on urban land. The use of farmland for urban purposes means a reduction in the land available for food production. In the 1990s, we could regularly read in the newspapers about how much farmland was lost due to urbanization and desertification. Taking into consideration the loss of land calculations could then be made to show how many kilos of rice per inhabitant was available. (Thinking of the irresponsible attitude and practice in my home country of reducing the land available for food production, this public information and transparency in China filled me with envy). At that time the Chinese policy was to strive for self-sufficiency of food. Later, when China increased its export industry, this policy was changed. But using as little land as possible for urban development has still been the main consideration. It is a consideration that is, of course, compromised by many exceptions, but at least it gives an argument for building concentrated, high-rise districts in the urban tissue. If the demographic trends are to be taken as a given, it seems rational to do so. The alternative is to release the pressure by decentralizing settlements, and this is an alternative that currently seems very far-fetched. To understand some of the reasons why it has all turned out this way, we need to see this modern housing development in light of urbanization and food supply. (When using the word understand, that does not, of course, necessarily mean accept).

Slum areas, as we find them in African and Latin American cities, were not to be found in China in the 1980s and 1990s. Seen through European glasses, the oldest, densely populated urban areas appeared to be slum districts, and were often characterized as such. And indeed, the dwellings were poor. But there was access to water, and there were public toilets and bathrooms, sewage systems, garbage handling, schools, and health stations in the neighborhood. Maybe something looked very shabby, but at least an infrastructure was there. Taking a closer look, one was impressed by how it was possible to live with some degree of dignity despite the simple conditions (which, by the way, is a striking feature of slum areas all over the world). Mind that this does not mean everything was idyllic and harmonious. In 1990, we interviewed people living in the central areas of Xi'an, named DTMD, where the average dwelling area was 2.7 square meters per person, and found that quarrels and clash of interests were common. A typical conflict would be when someone tried to grab a few square meters from the public space between the dwellings (see chapter 4 for a description of ingenious methods to do so). Scarcity of space had the result that people needed to relate to each other in different ways than what we were used to, while I am coming from a society with few people and an abundance of space around us.

The villages we have studied have changed in various ways. Some of them were dilapidated and semi-abandoned; some of them were in good condition. Some were in imbalance and seemed to be in doubt about whether they should continue to exist. Others chose to bet on two horses as they partly maintained old enterprises and partly invested in tourism. In all, the villages there were valuable historic buildings in decay, threatening to collapse.

Transport

In 2018, I prepared a tour for a small group of friends, intending to travel around China by train. We ordered tickets through the internet, on the national railway company's well-organized homepage. The order was rather complicated because there were many stops and breaks on the planned route, and also because there were different desires and time schedules among the participants. All this was properly handled by the case handler and payment was smoothly arranged. The stack of tickets was delivered to the hotel reception three days in advance of our arrival, (in a small hotel that hardly had any professional reception desk), and all tickets were in perfect order. The first leg of the journey, Beijing-Shanghai, was in comfortable seats on the high-speed train. What a change compared to the Hard Seat tickets on the steam engine train on our first travel inside China!

Parallel to the improvements of the train system, other long-distance transport infrastructures have also been built. Highways, bridges, and, not least, air transport installations. The limited number of air companies in the 1990s (one airline name, CAAC, was jokingly said to be an abbreviation for China Airways Always Cancelled–this was perhaps a slightly unfair characteristic, but it was also not too far from the truth) has now become a myriad of companies for domestic and international flights, with aircraft of high, international standard. New airports and terminals pop up continuously, often designed by famous architect companies from other countries.

The new urban landscape came with new traffic problems and a need for new transport systems. Housing and work units were originally organized close to the work enterprise, which was responsible for procuring housing for the workers. As soon as the Opening Up policy was introduced, housing was changed into a commodity in the housing market, and, as a consequence, people might suddenly live far away from their workplaces. This, in turn, created the need for motorized transport. The sheer size of the growing cities also demanded new transport solutions. In Xi'an in the 1990s, moving around by bike was the prevailing way of transport in the city. Ten years later, because of the heavy motorized traffic, it was dangerous to do so. This was not least because of the volume of private cars in the streets and the parking lots adjacent to the streets, even on the sidewalk. Horses and mules of the early 1990s were soon replaced by lorries. Noise and dust made their marks on the atmosphere of the street. To reduce these problems and to make urban transport more efficient, well-functioning metro systems have been developed. As is the case in so many big cities of the world, the underground transport system of the Chinese metropolises also lacks in capacity, but new metro lines are gradually added. In the old cities, there are often archaeological objects to be found 10-20 meters in depth, making it a puzzle to build tunnels without disturbing the underground cultural relics. In Xi'an, for example, this causes a considerable delay in the construction of metro lines.

Although they are now mostly replaced by regular car taxis, motorbike taxis and rickshaw taxis still exist. In the 1990s, taxis were numerous and easy to get. They competed to find the customers, sometimes causing accidents when unorthodox maneuvers were used, as they tried to beat each other to the customer. To save money, most people preferred the bus. In recent years, however, the scene has developed into a seller's market: During most of the day, one has to be very patient or fortunate to get hold of a taxi. But as apparently more people can afford taxis, buses are still the prevailing urban public means of transport – with overcrowded bus stops. Gradually the queue culture has changed. While earlier you would have to fight with elbows and knees to get in-

to a bus, queues are now better organized to not give prefer-
ence to the heaviest and strongest – and most impudent.

Pollution

Air pollution has always been a big problem in many re-
gions of China, due to the fact that dust from erosion in
the desert and plains in the north has been spread by wind
over the whole country. Sometimes hard storms spread this
type of dust even further south. It has been going on for
millions of years and will not end until the sandy land is not
sandy anymore, but green, which I believe will not happen
any time soon. Over years the dust has sometimes piled up
and settled, some places forming 200-300 meters deep
loess layers.

In addition to the pollution from air traffic, the effect on the
air quality from the increasing volume of private cars is substantial. Exhaust and dust whirled up by the traffic, combined with the smoke from polluting industry, gives health
problems which are a big headache for the political decision-makers. There was also in the early 1990s dusty air, but
this was, as described above, mainly caused by particles
blowing south from the deserts and plains of the north, as it
has been for millions of years. With man-made air pollution,
the conditions have worsened to a point where it is intolera-
ble. Even on a short stay, the heavy smog is a depressive experience. The difference between polluted days and days
of the clear sky – when it is possible to see the landscape
around the cities – is striking. Much of the Western media
has maintained that the Chinese government does not care
about this problem. That is, of course, not true. The fact is
that this is, like it is everywhere in the world, a big and very
difficult problem. A substantial contribution to a solution

would be to reduce the use of private cars in the cities. Some drastic and some less drastic measures are in place. They do have some effect because common people now understand the problem as they are personally experiencing respiratory problems. And maybe they also feel that being stuck in traffic jams for hours is not the freedom of movement that the auto industry promised.

The European cities are facing the same problem, albeit to a less conspicuous and comprehensive degree. When entering China from abroad – as the airplane dives into the smog over the big cities – the contrast has always made for a depressive experience. The Chinese authorities struggle by means of different measures (in traffic systems, energy use, and supply, closing polluting factories, modernizing domestic use of energy, etc) to improve the air quality. In recent years, I have personally experienced improvements; during the last two visits to Beijing, the sky has been clear, and it was possible to see the mountains to the north. Statistical data also show that the problem, though still very serious, is decreasing. In the global fight against the climate crisis, these are encouraging signs.

Food

During my first years visiting China, food – staple goods – was rationed. Families received food stamps, just like we did in Norway after World War II. The Chinese families used ration cards when they wanted to buy basic food items at an affordable price. Foreign visitors did not have this problem. Restaurant food was cheap and cooked according to Chinese food culture. But common people could not afford to eat in restaurants, so housewives and sometimes the menfolk used much time to find the cheapest offers on the vegetable markets. There are still very poor people who struggle to have enough to eat, but a majority of the city dwellers can find whatever they need in the shopping malls, at a price that seems reasonable for their family budget. Farmers in the rural districts may have less money to spend. I once interviewed a farmer family, and they maintained that they lacked the money to cover daily expenses, but because they produced food themselves, they always had something to eat. (I have never visited the really poor villages in remote locations where farming is harder, so I have not personally observed their problems. I certainly do not mean to imply that serious food shortage problems do not exist).

Sometimes hospitable people would invite me into their homes to share their food. It was always a pleasure to see how they made well-tasting and healthy meals, well composed and served in a friendly atmosphere. Visitors are usually invited to restaurants, either with a few friends or in a more official banquette setting. And then, too, it is a hospitable and friendly atmosphere. I have noticed one change in this type of meal. During the first years, the meals were overwhelming, both in quality and in the number of dishes. I learned that some foreigners even asked for simpler meals, but the hosts did not want to change this. However, in recent years the norm has changed (a change I warmly welcome) to simpler meals – this means fewer dishes served, less meat, and more vegetables. The reason for this was that the government, in an attempt to fight corruption, introduced regulations to limit the abundance and hence the costs of the official meals.

Trade

When no one is self-sufficient with everything (which is pretty much how it is for everybody everywhere), buying and

306

selling are necessary. How this is done has really changed in the last 30 to 40 years. During my first stays in China, foreigners received the special currency, FEC (Foreign Exchange Currency), which allowed the holder to shop in the so-called Friendship Stores. In these stores, one could buy things that were not for sale in regular shops. FEC money was very popular among local people because the assortment of merchandise was very, very limited in the shops to which they had access. Many people had money, but there was not much to spend it on. Now cities are like cities everywhere. The big cities have shops with a rich variety of commodities, and towns and villages have shops for daily use necessities. Shops address the streets with advertisements that change the visual image of the urban landscape. If you have money, you can get what you need.

In the early 1990s, there was only one bank in Xi'an accessible to foreigners, the Bank of China. This is where one had to go to change traveler's cheques into FEC money. For important transitions, one had to go to the Bank of China's headquarters in Beijing. Now the cities are full of bank buildings, which are usually the most pompous buildings in the city. And there are ATMs on every street corner.

Internet shopping is escalating. On a university campus, one can witness how students order anything with some clicks on their cell phones. A few days later the packages are brought on motorized tricycles and spread out on the sidewalk near the entrance to the campus. Students come there, fetch their parcels and try on clothes on the spot. If they are not satisfied, the parcel can be returned by the same people who brought it. It's high-tech trading combined with the low-tech distribution.

Internet trading has developed in recent years. More and more people shop online, and this has influenced the shopping infrastructure. A colleague of mine pointed to a rather new, 10-story building, and told me that a few years ago the whole building was a market for electronics. Now the building is empty, outcompeted by internet commerce. He himself also prefers to shop on the internet: Living outside the 5th Ring Road in Beijing, should he want to go to the city center, he would be stuck in several traffic jams on the way, and then he would have to find parking for the car downtown. A trip like that has become a full day's job. In his neighborhood, there are several nice restaurants. Most people chose to order food from these restaurants and to have it brought home to them by a bicycle errand boy.

Culture

Culture is a comprehensive notion and is used at all levels of life. I will just select a few issues that I feel describe characteristics of Chinese society. Partly because they, at first sight, seem to be different from my own culture, but by closer reflections may be something we have in common.

Hierarchical thinking exists everywhere. As a tool for orderly structuring, a hierarchy can be practical for arranging, categorizing, administering, planning, and implementing. In a huge state like China, it seems rational to base political practice on hierarchy. (This is also the case in other countries, even a small country like Norway with its 5-6 million inhabitants). In such hierarchies, it's a natural consequence that political power is executed along the same lines; how this works is depending on how communication and information move within and across the hierarchy, and how influence from the outside can enter into the different levels. A

Chinese colleague pointed out to me how, also outside the political sphere, hierarchical thinking is integrated into people's life. As an example he mentioned that in China, when one writes an address on a letter, one starts with the country, then the province, then the city, then the street, then the work unit, and finally the name of the person who should receive the letter. In Western countries, we do it the other way around. We write the name of the addressee at the top, then the work unit, then the street address, the city, and then lastly the name of the country. Does this reveal totally different ways of thinking? If we understand culture as what is in the mind of a person and is also shared by the other members of the society, can we then say that hierarchical thinking is part of Chinese culture? Perhaps it's a stretch to describe the Chinese and Western cultures as fundamentally different based on this example–there are so many exceptions blurring the idea. But I nevertheless feel that the statement is valid. As far as I could see, this is unchanged, maybe the most stable of the cultural features, and it penetrates all kinds of communication I experienced during the few decades I have been in China. How far back do the roots of hierarchical thinking stretch? Maybe 2,500 years, to the time of Confucius. Maybe even further? I won't insist on a hypothesis, merely maintain that this feature of Chinese culture has been evident to me in the time I have communicated with friends and colleagues in China.

Culture is, in any country, used for political and commercial purposes. It may be employed in the manner of what we call "soft power." It was used in this way, for instance, in the Belt and Road Initiative initiated in 2013, where the original Silk Road has been a backbone of the project known to all Europeans from the adventures of Marco Polo. The Confucius Institutes, initiated in 2004, is another example drawing on the vague knowledge we all have about the philosopher Confucius, who lived from 551 to 479 BC. Confucius institutes spread information about China and are now well established worldwide. And like in all countries, culture is used for domestic, and political purposes, a tool for a continuous identity formation of the country and its inhabitants. While using identity as a constructive element for the development of society, it seems hard to find a balance between national pride and national chauvinism. The establishment of the two archaeological parks in Xi'an, the Tang Dynasty Daming Palace, and the Han Dynasty Han Chang'an City (see chapter 10), are examples of where the national government has made huge financial and political investments in order to show the glorious past of the country. This has been important for attracting tourists to the city, and it has been important to create pride in history among local and domestic visitors.

Xi'an, the capital of 13 dynasties, along with the neighboring districts, have numerous cultural relics and sites, where there have been archaeological excavations since the 1950s. The famous Terracotta Warriors were found by accident in 1974. Since then, museums and ways to visit the site have been developed for domestic and foreign tourists. When I first went to see the Terracotta Warriors in 1989, they were exhibited with uncomplicated and simple museology. Later the presentations were more and more elaborate, in order to satisfy the escalating tourist industry and professional demands for protecting the cultural heritage. Back then I went to the site by bike on bumpy local roads. Nowadays visitors go there by bus or taxis on four-lane highways. There have been additional excavations and numerous, new finds. New monumental buildings have been raised around sites, underlining the importance of having been enrolled in the WHL.

The millions of tourists who visit every year are welcomed by modern facilities.

Similar developments have occurred in other historic cities. It is noticeable that already many years ago the number of domestic tourists surpassed the number of foreigners. Encouraged by the government to see their own country, they join the long queues to enter the most famous spots. When, for example, I visited Beijing in April 2018, the wait in line to enter the Forbidden City was an hour, and this was despite a very efficient ticket-sale system and control. In tourist season, like the holidays around the National day in October, over-crowding is excessive – to great frustration, of course, for the many who travel far and want to make the most of their holidays and instead end up spending most of their vacation waiting in lines. This experience is quite different from when I first traveled to China, at a time when domestic tourists were few and foreign tourists were a sensation, and the gazing tourist was very often met with a gaze right back.

The development of the Qujiang District in Xi'an, which is the area surrounding the Big Wild Goose Pagoda, represents new models for urban development. It uses cultural monuments and history to promote investments and urban life. Restored several times, the Big Wild Goose Pagoda is a Tang Dynasty monument. In 1985, it was surrounded by the farmland areas south of the city, adjacent to the Temple of Great Maternal Grace. I went there by bike through a pastoral and peaceful landscape, and there were no tourists there when I reached the pagoda area. Today the district is a thriving, urban milieu with restaurants, cinemas, museums, shopping centers, theaters, parks, and facilities for the public, including a monorail track high above the ground. The district is a result of a special planning approach. The city initiated the development of the area by first investing in the necessary infrastructure, and only when this was in place, utilized history as a marketing tool to attract developers to the district. The Big Wild Goose Pagoda is, of course, a major component, but to supplement the historical references there are also a few archaeological relics exposed in ditches, myriads of new, stone-carved poems from old times, and monumental sculptures picturing historical persons of importance. Qujiang District is a recent experiment on an urban scale in Xi'an where history and commercial interests are very closely linked.

One of the reasons why it is so interesting to visit China is the mix of old traditions and the expressions of modern times. The scripture system, for example, which for a foreigner has exotic, visual qualities, is for a Chinese an important part of the national identity. I've been told there are more than 80,000 characters, and clever people have managed to make it possible to write and read those characters on PCs and mobile phones, managing to write very fast with all those characters. Compare this with the Scandinavian languages which have a 29 letters alphabet. The English alphabet has 26 letters, and the three extra Scandinavian letters, æ, ø and å are still not fully introduced into the digital world. What a contrast this is to the ultrafast modernization of Chinese culture!

刘克成、肖莉访谈：
一位挪威建筑学者的人生启示课

时间：
2021年5月20日
地点：
中国西安、
挪威特隆赫姆
受访人：
刘克成、肖莉
采访人：
王宇、许东明

受访人简介：

LIFE LESSONS FROM
A NORWEGIAN PROFESSOR

INTERVIEW WITH
LIU KECHENG AND XIAO LI

Date:
Thursday, May 20, 2021
Place:
Xi'an, China, and Trondheim, Norway
Interviewees:
Liu Kecheng and Xiao Li
Interviewer:
Wang Yu, Xu Dongming

刘克成（1963— ），西安建筑科技大学建筑学院教授、陕西省古迹遗址保护工程技术研究中心主任，中国当代知名建筑师。长期从事与文化遗产保护相关研究与规划、建筑设计，曾主持汉阳陵、秦始皇陵、唐大明宫国家考古遗址公园等保护规划；主要建筑设计作品包括汉阳陵遗址博物馆、大唐西市博物馆、秦始皇陵百戏俑遗址博物馆、文吏俑遗址博物馆、西安碑林石刻艺术馆、三门峡虢国君王墓地遗址博物馆、南京金陵美术馆、南京中国科举博物馆等。

肖莉（1959— ），西安建筑科技大学教授、建筑师。长期从事建筑设计及理论领域教学、研究与设计工作，主持及参与项目曾获国家优质工程银质奖、全国优秀勘察设计二等奖、中国建筑学会建筑创作奖、世界建筑师协会遗产保护大奖等多项荣誉。

采访人：您和哈罗德教授是怎样开始交往的？能否谈谈您对他印象最深的几件事？

刘克成：回顾我跟哈罗德的交往，最早应该是1995年。1996年，我开始在西安建筑科技大学建筑学院做副院长，负责科研。在我接副院长前一年，哈罗德教授、法国的让 - 保罗（Jean-Paul Loubes）教授，还有其他几位教授，打算在西安开一个关于西安民居方面的国际会议。那时哈罗德先生在西安交通大学和董卫老师在西安北院门做了很多工作。之前这个会议主要是去法国的几位老师联系的，当时约定的就是中、法、挪三国学者围绕着西安民居话题开一次国际会议，但是不知什么原因，人都到了西安了，这件事儿没人管了。我被派去灭火，那是我第一次见到哈罗德先生。

让我印象最深刻的，就是他思路极其清晰，讲事情组织条理非常清晰，和我们谈了中、法、挪三国这个国际会议怎么办的问题。法国的让 - 保罗教授非常生

Interviewer: Let's start with a general question. How did you and Professor Harald Høyem get to know each other? Can you tell us some of the things that impressed you most about him?

Liu Kecheng: We first met each other in 1995. And in 1996, I started to work at Xi'an University of Architecture and Technology as Associate Dean in charge of academic research in the architectural school. In 1995, a year before I became the associate dean, Professor Harald, Professor Jean-Paul Loubes from France, and some other professors were planning to hold an international conference in Xi'an on traditional housing in the city. By that time, Harald had already done a lot of work in Xi'an Beiyuanmen Street in the DTMD with Professor Dong Wei, who worked at Xi'an Jiaotong University. This international conference was initiated in the first place by some Chinese scholars who had a connection with France, so most of the participants were from China, France, and Norway. However, for some reason, no one was taking responsibility for organizing the conference when all the participants had already arrived in Xi'an. I was then sent to put out the fire, and that was the first time I met Harald.

What impressed me the most was that when we talked about the arrangement of the international conference, Professor Harald was very organized and clear-thinking. Professor Jean-Paul Loubes from France was angry about the situation and all I could do was to apologize on behalf of the School of Architecture and promise more organized and effective future collaborations. Thus, my first contact with Professor Harald occurred under less-than-ideal circumstances. Nev-

314

气，说了很多，我只能代表建筑学院表示抱歉，然后说这个事情我们怎么往前推进。这是我们交往的开端，其实并不是在一个特别好的气氛下开始的。我印象是这样。作为负有责任的一方，我是从道歉开始的。但是哈罗德先生给我的一个非常深刻的印象，是自始至终保持一种特别温和的态度，只是在谈事情，谈怎么来推进三国的合作，怎么来推进科研，这让我印象非常深刻，就像遇到一位老父亲，以一种宽厚温和的长者态度来面对这件事情。

这之后断断续续几年中，因为在建筑学院负责科研这块，我也出了一些点子，把西安碑林还有其他一些课题推介给他们，让大家一起带学生做了一些事情。到1998年，就是两年后，受哈罗德先生邀请，我第一次到挪威。应该是元月，我印象中是大雪天，法国的两位教授也同时到挪威，谈进一步合作的事情。大家就住

在哈罗德家里，我住在他那个小阁楼的二楼。虽然之前去欧洲很多次了，但是到挪威仍然是一个非常新鲜的感觉。哈罗德来接我，是一辆很老式的汽车，开到他家门口。那是冬天元月，大雪封山，车停下来，距离门口还有好几米远。然后我们踏着雪走到门口，门口放了一个木平台，哈罗德先跺跺脚，让我也跺跺脚，再进去。我的第一个问题就问他：为什么从道路到家跟前，你还要一片自然的草地？为什么不在这儿弄一块儿硬质铺面？老先生非常认真地跟我说，跺跺脚这件事并不费事儿，但是如果多做了一块儿硬质铺面，对环境的改变就很多了，多这一平米是没有必要的。这让我真的印象深刻，我所了解挪威人对自然的态度，应该说就是从哈罗德家的门前上的第一课。

那天到的时候已经接近深夜，第二天早上起来，出于建筑师的本能，我围

ertheless, I was very impressed by Professor Harald's gentle manner throughout the whole process, whether I was apologizing or we were discussing how to promote further cooperation among the three countries. When I talked with Professor Harald, it was like talking to a generous and easy-going senior member of my own family.

In the following years as the associate dean, I also came up with some ideas and recommended some projects, including the Beilin project, to Professor Harald. We have completed some projects and research together with students from China and Norway. In 1998, I was invited to Norway for the first time by Harald. I think it was in January, I remember it was a snowy day, and two professors from France came to Norway at the same time. I stayed in the west loft of Harald's house. Although I had been

to Europe several times before, it was still a fresh feeling to be in Norway. Harald picked me up at the airport in his old car. When we arrived at his home, he parked the car and we walked to the door which was about ten meters from the garage. There was a lot of snow on the ground. Before we went inside the house, Harald stomped the snow off his shoes and so did I. I remembered clearly that my first question to him was "why didn't you make a pavement path to your doorstep?" and his answer was "then I would have to change the environment around the house. It is not a big deal to stomp the snow off my shoes." It really impressed me. Thus, I learned my first lesson about a Norwegian's attitude towards nature at the doorstep of Harald's house.

It was almost midnight when I arrived that day, and when I got up the next

着房子周边认真转了转，哈罗德陪着我，这时才看到整个的房子——他那个住宅大部分是支在几个很小的混凝土柱基上（图16-1）。*1 哈罗德跟我讲，当年这块地是他父母买的，他年轻时候设计盖的。为什么建筑结构不全部落下来，而是支在几个点上？他说也许以后子女还喜欢住这儿就继续住下去，如果有一天子女不喜欢，这房子也许会拆掉，拆除后对这个山坡的自然环境改变也会很小。

　　1996年那一年，我起草申请了关于绿色建筑的全国自然科学基金的重点项目，这件事是刘加平老师他们去做的。为这件事当时花了很多时间研究绿色建筑的有关国际文献，因此我对绿色建筑是比

图 .16-1
哈罗德自宅冬景
（摄影：许东明）

Figure.16-1
Harald's house in winter (Photo by D.M.Xu)

*1 其住宅为20世纪70年代初哈罗德在丹麦做职业建筑师时设计自建，局部以混凝土砌块做护坡形成地下室，大部外墙置于独立柱基之上。

morning, out of my architect's instinct, I took a serious look around the house (Fig. 16.1) accompanied by Harald. The foundation of the house was built on top of some small concrete pillars. Harald told me that the land was bought by his parents and that he had designed and built the house in the 1970s when he was young. I was wondering why the foundation was built on pillars. He said that his children may continue to live here if they wish, but if they don't, the house will be torn down and the natural environment of the hillside will have been minimally changed after the demolition.

In 1996, I drafted an application for a key project of the National Natural Science Foundation on green building, and Professor Liu Jiaping and his team carried out the project. I spent a lot of time studying international literature on green building,

so I am quite familiar with the topic. In fact, the concept of sustainable development was introduced by former Norwegian Prime Minister Gro Harlem Brundtland in 1987 and became a global agenda at the Rio Conference in 1992. But I have to say that these two lessons given to me by Harald were a real eye-opener. I mean, in a country like Norway, which is one of the most developed and livable countries in the world, Professor Harald used his daily life experience, the house he built, to tell me what sustainable development should be: green building does not necessarily mean high building cost and high technology. It should start with self-consciousness and self-restraint, which is one of the simplest methods. What Harald taught me is quite different from what I have talked with some experts in the US and Germany about regarding green

316

较熟悉的。其实"可持续发展"这个概念，就是1987年挪威前首相布伦特兰（Gro Harlem Brundtland）女士提出的，到1992年里约热内卢会议成为一个全球性的纲领。但是听了哈罗德给我上的这两堂课，真的是别开生面。在挪威这样一个今天世界生活品质最高的国家之一，哈罗德教授以他的日常生活、他建造的住宅来教育我，可持续发展应该是什么。就是绿色建筑体系并不必需要花很多钱，也并不必需要是一个高科技的东西。其实很重要的是，从人类的自我约束开始，这是一个最朴素的方法。这跟之前我接触到的相当多的美国人、德国人所谈的绿色建筑和可持续发展都是非常不同的。这事到今天为止都让我记忆深刻，我觉得这是我初到挪威哈罗德先生给我上的两堂最好的课。

第三课呢，仍然是第一次到挪威时。哈罗德住在山上，在大雪天的时候，上山下山都不是一件很容易的事儿。我印象挺深的，就是他每次开车把我带到挪威科技大学校园边上一个大草坡，车停在底下，然后人再步行上去。有一次我在学校待着，他从家里过来，在雪天骑自行车到学校，这让我也挺吃惊的。因为在挪威这样一个发达国家，开车我觉得是一个自然的选择。当他骑着自行车背着双肩包来上课的时候，那天中午跟他吃饭，我就问他这是什么意思，他说这很自然，也可以少烧点儿汽油，少点儿污染。他告诉我在多数情况下他不太用汽车，主要就是骑自行车，有时时间充裕，步行回家其实也有意思。我印象中我也有那么几次是步行回家的。在我去德国、法国以及美国这些其他国家认识的教授，还很少见到像哈罗德这样的，总有这么强的自律。我觉得"可持续"这个概念，在他脑海里面是一种自我生活的原则，这真的是让我特别敬佩的事情。

buildings and sustainability. These are the most meaningful lessons Harald had given me at the start of our relationship.

The third lesson I got from Harald was also in Norway. Harald's house is located at the top of a hill, and it was not so easy to access when there was heavy snow. On one such day, I was at the NTNU Gløshaugen campus and found Harald had ridden his bike there. That really shocked me because I assumed that driving would be the natural option. I have seen several times Harald rode his bike to the campus with a backpack on his shoulders. I asked him why, and his answer impressed me. He told me that cycling is a natural choice because it is more environmentally friendly and that he rarely drives. When it is good weather, it is enjoyable to walk home as well, he said. After hearing that, I tried walking several times from campus to Harald's house. Among the professors I have known in Germany, France, and other countries like the United States, I have hardly met one like Harald, who always had such strong self-discipline. I think the concept of sustainability is a living principle in his mind, which is something that I particularly admire.

The fourth lesson Harald gave to me was in China. In 2000, Harald applied for more than 4.6 million NOK from Norwegian Agency for Development Cooperation for the conservation and maintenance of three selected courtyards in Beiyunmen District in Xi'an. Harald had the authority to allocate the fund as the project leader. He always chose to live at the cheap university guest hotel and to take the bus most of the time even though it was not costly to take a taxi in Xi'an at that time. My Chinese col-

第四课呢，就到中国了。2000年前后，哈罗德先生从挪威外交部发展合作署（NORAD）申请了460多万元，来做西安北院门三套院子的保护维修。按照基金要求来说，这笔钱他有签字权，他签了字就能够使用了。他住在学校宾馆，那个时候，西安的出租车是很便宜的，五块钱起步，打个车到钟鼓楼，不会超过十块钱。在这个项目执行期间，坐出租车从学校到北院门应该不超过十块钱。而大多数情况下，哈罗德去现场都是坐公交车去。你们也知道，其实从学校坐公交车到那里，至少要倒一次车，并不是很方便。对中国人来说，我自己有时候都偷懒，打辆车就过去了。但是哈罗德只要是自己去，每次基本上都是乘公交车去。对于这个项目来说，我认为打个车属于正常开支，并不过分。那时候打辆出租车也很容易。但他就是以一种极其自律的方式，只要是自己过去都

是坐公交车去现场。让我觉得特别敬佩。

第五课呢，是在挪威北部的世界遗产地维加群岛（Vegaøyan），那个岛也是一处知名的海鸟栖息地。*2 那年（2005年）哈罗德组织活动，不只是我们，还有北京中规院的王景慧先生、清华的张杰老师和西安这边一批人，大家一起去那个岛上。我比较喜欢捡石头，走到某处，见到有意思的石头会揣到口袋里带回来。那岛上的石头非常特别，上面有白色的斑纹，像凝固物一样，我就捡了几块石头放在口袋里，然后我还有点儿炫耀地拿给哈罗德看，说看这个石头多漂亮。哈罗德很温和

*2 维加群岛（Vegaøyan）邻近北极圈，包括约6500个小岛，其主岛自石器时代即有人类居住，岛上居民至今仍然保留了渔业、农牧业等传统特色生活方式。此外，维加群岛也是国际鸟盟（Bird Life International）确认的重要鸟类栖息地。2004年，维加群岛被联合国教科文组织列入世界遗产名录。

leagues and I often took taxis, and for this project, taking a taxi would not have been an excessive cost. But still, Harald was very self-disciplined, and he would not choose a taxi if buses were available. In contrast, some of the Chinese partners in this project often treated themselves to expensive meals, spending a lot of money on one meal when they received the funding. I was quite ashamed of such a performance.

The fifth lesson Harald gave to me was at Vega Archipelago in Norway. Vegaøyan is a World Heritage site in northern Norway and is well-known for its habitat for seabirds. That year (2005), it was not only my colleagues and me but Professor Wang Jinghui from the China Academy of Urban Planning and Design and Professor Zhang Jie from Tsinghua University who were invited to visit the island. You all know

that I love collecting stones. The rocks and stones on that island were very special, with white flecks on the surface. I picked up a few of them, put into my pocket, and showed them to Harald with a little flair. I will never forget what Harald said to me. His tone was gentle but sounded serious to me. He said, these rocks are certainly beautiful, but this is a world heritage site. If each of us picks up a few of them, then no one else will be able to see them in the future. I felt ashamed and quickly put the stones back. After this, when I went to heritage sites or even non-heritage sites, I started to feel somewhat self-conscious. It is true that you cannot take everything just because you like it, but you also have to think about what the consequences would be if everyone does the same thing as you.

In the decades Harald and I have

地说了几句话，其实我觉得挺严厉的。他讲这个石头当然很漂亮，但这是世界遗产地，如果我们每个人都这样捡几块石头，那以后别人也都看不到了。我觉得很羞愧，赶快把那几块石头放了回去。这以后，当我到遗产地的时候，甚至非遗产地的时候，我开始有种自觉。很多东西确实不是说你喜欢，觉得漂亮就可以把它拿走，你还要想想，当很多人都这么做的时候，会对遗产地造成什么影响。

　　　　我和哈罗德相处几十年，经历了很多大大小小的事情，有在中国发生的，有在挪威发生的，也有在其他地方发生的。我觉得哈罗德对我来说，他始终有点儿接近一个父亲的形象。其实就我跟我父亲的关系来说，成年以后，自己跟父亲的交流，不算太通畅。他也不是干这个专业，彼此共同话题并不多。但从我跟哈罗德的交往来说，我觉得他真的有点儿像文学作品中

描述的那种理想的父亲形象。一个父亲对孩子，如果能够像哈罗德这样，我觉得就是一个非常完美的父亲形象，特别有良心，有个人操守，一个真正可以大写的人。

采访人：确实如此。我们作为晚辈和旁观者，对您和哈罗德先生这么多年交往的很多细节也印象很深。比如您是典型建筑师生活作息，有熬夜工作晚睡晚起的习惯。但每次哈罗德离开西安启程回国，即使时间再早，您也一定会早早起身过来亲自送行。今天听您说这些，多少能更好理解你们何以能有这样一种交谊深厚的关系。

　　　　按哈罗德先生早前对他这本中国回忆录的设想，他是想请您以中国建筑师看挪威的视角专门写一个章节，具体谈谈您对他此前在中国这些年所做工作的观感。刚才您也谈到了哈罗德及其挪威同事团队在西安北院门完成的传统院落保护项目。这三套院落也是今天西安明城区内硕果仅存、为数不多的传统院落，不知您对这个项目怎么看？

been together, we have experienced many things together, in China, Norway, and elsewhere. I feel that Harald is sort of a father figure for me. In fact, as far as my own relationship with my father is concerned, after I became an adult, I did not have much communication with him. He is not a professional in this field either, so we do not have much in common. But it is true that from my interactions with Harald, I really think he is like the ideal father figure described in the literature.

Interviewer: Impressive. In all those years, we have also witnessed your friendship with Harald and we were also impressed by some small details. You are a typical architect who has the rhythm of staying up late to work and getting up late in the morning. But every time Harald left for Norway, you

would always get up early and see him off in person. According to Harald's idea for his memoir about China, he wanted to ask you to contribute a chapter about your perception of his previous works in China from the perspective of a Chinese architect. You also talked about the traditional courtyard conservation project that Harald and his Norwegian colleagues completed in Beiyuanmen District in Xi'an. These three courtyards are one of the few remaining traditional courtyards in the Ming Dynasty city wall of Xi'an today, and what do you think of this project?

Liu Kecheng: The courtyards in Beiyuanmen he preserved are undoubtedly very precious assets. In fact, from the late 1990s to around 2000, we conducted fieldwork on traditional houses for the Cultural Heritage Administration of Xi'an. At that time,

刘克成：北院门的这几个院落，无疑随着时间的推移越发显现出它们的珍贵。其实在20世纪90年代末到2000年左右的时候，我们中心替西安市文物局做传统民居调查，那时在西安城内各个街巷里面，实际上是保留了至少160套左右比较完整的传统院落。到我们去年再统计，剩下的已经不足30套。换句话说，在这20年中，实际上已经所剩无几了。而在这所剩无几的院落之中，哈罗德先生曾参与保护的三套院子，是其中质量最好的。因此确实可以这样说，随着时间的推移，我觉得这更显示出哈罗德先生和挪威同行在做这件事情上的远见和卓越，这是我想要强调的第一点。

第二点，我觉得在这个工作过程中，哈罗德先生以及挪威团队所采用的一些工作方法，无论是作为学者还是建筑师，对我们来说都受益良多。像哈罗德先生、女教授艾尔（Eir Grytli），还有达格（Dag Kittang）教授，包括专门做GIS的博泰耶（Bo Terje Kalsaas）等好几位教授，我觉得是很有意思的一个团队组合（图16-2）。他们非常重视基础资料的建设，实际上挪威外交部发展合作署这个项目基金所投入的钱里面，有相当大的比重，是在做基础资料建设。这是让我感觉到我们与发达国家成熟研究团体最大的一个区别。我们经常项目就是项目，很少考虑这个项目结束以后，后面的人怎么来接的问题。而挪威团队一开始实际上是做基础数据库，从GIS系统数据库的建立开始，而且花了大量时间精力做这件事。这个工作我们以前虽然也在做，这也是当时中方团队肖莉老师投入精力最大的一个方面，但是从来没有用GIS系统这种大数据的方式来做。这是第二个我觉得非常受益之处。但特别遗憾的是，因为当时选择的中方对接单位是

at least 160 traditional courtyards were left intact in various streets and alleys in Xi'an. However, the investigation we did last year showed that there were only less than 30 courtyards remaining. In other words, during the past two decades, the vast majority of them have disappeared. Among the few remaining courtyards, the three courtyards that Professor Harald and his Norwegian colleagues worked on are the best-preserved ones. So, I would say that it shows the importance of the vision of Professor Harald and his teammates with the passage of time.

Furthermore, we have learned a lot, both as researchers and architects, from the working methods Professor Harald and his team have adopted. They had formed a very impressive team with Professor Eir Grytli, Professor Dag Kittang from NTNU, Professor Bo Terje Kalsaas, and so on (Fig. 16.2). Data collection was of great importance in their work, and much of the funds from NORAD were used for data collection. This is one of the biggest differences between us and the mature research community in developed countries. When we are doing a project, we just focus on the project, not thinking of what we could leave for future research. The Norwegian team began with the data collection at the very beginning by building the GIS and putting a great effort into it. Although we had done similar data collection before, we had never used an advanced system like GIS. It is a pity that all the collected data disappeared when the people in charge of the project left. They were working at the Xi'an Municipal Planning Institute under the Xi'an Municipal Construction Committee, the formal

西安市建委下的市规划院，那个数据库的管理也放在他们那里，随着具体做这件事情的人的离开，那套数据连同设备软件都消失了，这是一件非常遗憾的事情。

第三点，我印象是博泰耶教授让我们把项目中用的每块砖、每袋水泥的价钱都要统计出来，以及运输费、运输工具，等等。就是要把建造全过程中的能源损耗记录下来。在我的经历中，这是第一次做这样的事儿，当时对于我们也是一个全新的方法。因为在那个时候，我们还不知道可持续发展或者低碳减排研究这件事情到底怎么去做。

包括像北院门专题教学中挪合作课程，每次挪威教师上课，都是请莉丝贝特（Lisbet Sauarlia）先从社会人类学开讲，这个对我也触动很大。我们看场地，经常注重的是物质的东西，但是以人类学视角来看，其实首先注重的是人与社会的问题。我认为这是一个非常端正的视角，这也是今天

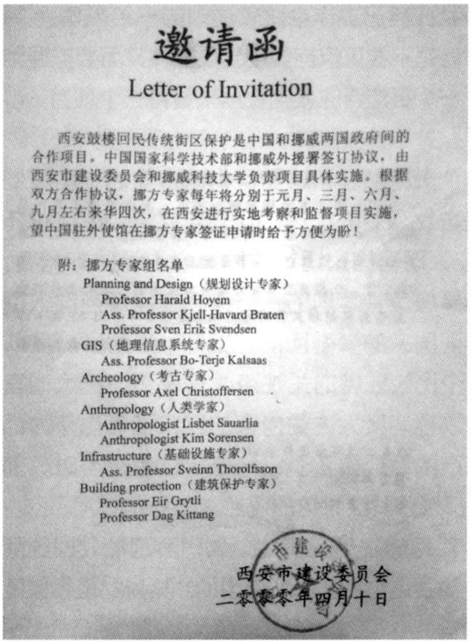
邀请函

Letter of Invitation

西安鼓楼回民传统街区保护是中国和挪威两国政府间的合作项目。中国国家科学技术部和挪威外援署签订协议，由西安市建设委员会和挪威科技大学负责项目具体实施。根据双方合作协议，挪方专家每年将分别于元月、三月、六月、九月左右来华四次，在西安进行实地考察和监督项目实施，望中国驻外使馆在挪方专家签证申请时给予方便为盼！

附：挪方专家组名单

Planning and Design（规划设计专家）
Professor Harald Hoyem
Ass. Professor Kjell-Havard Braten
Professor Sven Erik Svendsen
GIS（地理信息系统专家）
Ass. Professor Bo-Terje Kalsaas
Archeology（考古专家）
Professor Axel Christoffersen
Anthropology（人类学家）
Anthropologist Lisbet Sauarlia
Anthropologist Kim Sorensen
Infrastructure（基础设施专家）
Ass. Professor Sveinn Thorolfsson
Building protection（建筑保护专家）
Professor Eir Grytli
Professor Dag Kittang

西安市建设委员会
二零零零年四月十日

图 .16-2 西安市建设委员会为挪威合作团队签署邀请函影本（摄影：车通）

Figure.16-2 Invitation letter to the Norwegian team from the Xi'an Municipal Construction Committee (Photo by Che Tong)

Chinese partner of the project.

Another thing that has impressed me is that Professor Bo Terje Kalsaas asked us to keep a log of every expense, including the price of each brick and bag of cement, transportation expense, and so on. That was the first time I have ever kept a log in such detail because at that time we still didn't know how to do sustainable development or low-carbon emission reduction research. We had a Sino-Norwegian joint course on the topic of the Beiyuanmen project, and every time the Norwegian teaching team would begin the course with Associate Professor Lisbet Sauarlia's lecture on social anthropology. This was something new for us as well. We were used to focusing on tangible things from an architectural perspective, but from an anthropological perspective, the first thing we focus on should be the people and society. I learned a lot from this experience and that is the way I am working today - focusing both on the tangible and intangible aspects.

Interviewer: Among the projects Professor Harald and you have worked on together, the international concept design competition on Xi'an Daming Palace National Heritage Site Park from October 2007 to January 2008 must be mentioned. You were one of the people who initiated the compe-

我在自己工作中比较注重的一种视角——就是不要仅仅拘泥于物质实体，而要扩展到一个更广泛的人类视角来看待一个项目。

采访人：今天回顾您和哈罗德先生合作的一些重要项目，印象比较深的还有2007年10月至2008年1月的西安大明宫国家遗址公园概念设计国际竞赛。这件事也是您负责筹划的。记得您在点评当时入围的八个不同国家的参赛团队方案时，提到挪威团队方案中一个比较独特的关注点：即是否应把大明宫遗址区内以棚户区为主要内容的日常建筑也作为西安城市遗产的组成内容予以考虑？这也是当时其他参赛团队都未曾关注讨论的问题。*3

刘克成：确实如此。大明宫遗址公园国际竞赛这个点子是我们出的，其目的是希望让

*3 见本书第十章相关介绍及参见许东明.《挪威价值——基于大明宫国际竞赛挪威方案的遗产保护史与保护价值理念探讨》[J].《住区》，总第61期，2014 (3):17-23。

大明宫成为一个中国国内和国际古迹遗址保护的新范例。当时的想法就是这样。怎么达成这个目的呢？就是把国际古迹理事会（ICOMOS）的主席、副主席、司库这些人都拉进组委会，利用他们的影响力找一些全世界做遗址保护最出色的团队来参加，这是其一。第二是我们当时有一个比较明确的想法，就是希望邀请在遗产保护方面属于不同学术思想、流派，同时又有丰富实践经验的团队来参赛。所以最后有意大利团队，有日本团队，就是做京都奈良保护的那家，还有以色列、美国、澳大利亚和挪威等几个国家的团队。

对日常建筑遗产的关注，确实是挪威团队方案的一个很突出的特征。就是把历史当作一个连续的过程，而不只是保留一个切面、一个断面。不是只去看帝王将相，或者只看这个历史遗迹点最辉煌的故事，而是把它当作一个完整的历史过程

tition. I remembered that eight teams from different countries participated in it. When you commented on the different design concepts, I remembered clearly that you mentioned one thing that made the Norwegian team stand out. That is, the Norwegian team came up with one question which was not brought up by other participating teams: should the buildings in the shantytown in the Daming Palace National Heritage site be taken as a part of the city heritage?
Liu Kecheng: It was we who initiated the international concept design competition for Xi'an Daming Palace National Heritage Site Park. The aim was to make the Daming Palace case a new example of heritage conservation both in China and internationally. To achieve that, we invited the president and vice-president of the ICOMOS to sit on the competition committee and asked them

to help us find outstanding conservation teams to participate in the competition. We also wanted to invite teams that have different academic ideas on heritage conservation to achieve diversity in the competition, so there were teams from Japan, Israel, the United States, Australia, Italy, and Norway.

The focus on informal architectural heritage was indeed a very prominent feature of the Norwegian team. They see history as a coherent process, not just taking a piece out of history. Their focus is not on a particular and glorious moment of history, but on the whole history as an integrated piece. This is what we are talking about today in heritage conservation: authenticity and integrity.

This has always been the main principle for Harald and his Norwegian colleagues. We can see that in all the proj-

来看。其实在今天的遗产保护来说，这就是怎么正确地理解所谓真实性与完整性的问题。

我觉得这个也可以说是哈罗德教授和挪威团队一以贯之的一种视角。不仅是在大明宫这个项目，包括在北院门项目中对当地住民的尊重。在汉长安城遗址保护项目，其实对于这个问题的讨论更多——政府老想搬迁村子，哈罗德先生总是想着能把这个村子留下来。其实他对我在这个问题上的观点的改变也很多。后来在大明宫遗址公园实施的时候，我们也极力想保留道北那片棚户区，特别是丹凤门和含元殿遗址之间那部分。我当时还专门指定了两个研究生对那片区域做了仔细的记录，试图说服各个方面，能够把它保留下来。非常遗憾的是，对于保留这片内容反对最多的还是文物专家，因此最后还是拆了。不过在大福殿和麟德殿遗址之间，

我们留了18个院子。也可以告诉你们一个好消息：在经过十年的时间以后，现在我们终于说服对方，要把这18个院子永久保留，改造融入到大明宫遗址公园里面。到今天，这个目标终于有机会能实现了。

采访人：刚才您也提到汉长安城未央宫遗址这个项目。此次哈罗德在他的中国回忆录中也有专章讨论这个项目，*4包括他夫人玛丽娅（Marie Louise Anker）作为 ICOMOS 世遗申报项目评估顾问就此所写的专文，都表示不能认同把遗址区内的村子全部迁出这种做法。这处遗址在2014年也被成功列入了联合国教科文组织的世界遗产名录，至少说明针对这一遗址的整体规划和保护设计工作还是体现了较高的一个水准。不知您怎么看这种矛盾性？

刘克成：这个问题以我今天的观点来说涉及几个方面。从理论角度来说，怎么认识真

*4 详见本书第十章。

ects they had been involved in, not just the Daming Palace Project. We can see that in their respect for the local residents in the Beiyuanmen Project. I also remember in the conservation project of Han Chang'an City remains, we had a lot of discussion about authenticity and integrity because Harald and his team wanted to keep the village as it was but the authorities intended to relocate the village. Harald has indeed changed a lot of my views on this issue. Later on, in the Daming Palace Heritage Site Park Project, we also tried also very hard to keep the shantytown on the north side of the railway, especially the part between the Danfeng Gate ruins and the Hanyuan Hall site. I also asked two postgraduate students to take detailed documentation on that area, trying to convince various parties that it could be preserved. Unfortunately, it was the Chinese heritage experts who were most opposed to the preservation of the shantytown, so it was eventually torn down. The good thing was that after our continued efforts over ten years, eighteen courtyards located between the ruins of Dafu Hall and Linde Hall were kept as a part of the Daming Palace Heritage Site Park.

Interviewer: In Harald's memoirs about China, there is one chapter about the Weiyang Palace site in Han Chang'an City project. Harald's wife Marie Lousie Anker, who sat as the consultant of the ICOMOS world heritage assessment team, has also written an article about this. She expressed her argument for the removal of the villages from the heritage site. In 2014, the Weiyang Place heritage site was listed as UNESCO World Heritage, which is considered a suc-

实性和完整性？由于学者的立场、态度不同，在这个问题上的判断差别很大。我工作中经常吵架的对象都是考古学者。很多考古学者比较容易去选择一个历史断面，对它有特殊偏爱，认为为了这个历史断面而毁掉其他断面好像是当然的。你是强调遗址整个历史的一个完整过程，还是只强调某一个时间和空间断面上的真实与完整？我认为这个问题有很强的争议性，并不是说已经是常识。甚至非常著名的学者，在这个问题上的分歧都非常大。

　　针对汉长安城遗址的保护，那时正好是未央宫前殿遗址申遗，总负责人是陈同滨老师。我也把哈罗德就村子和住民的意见比较完整地跟陈老师谈过。而哈罗德也想和陈同滨见面谈谈，当时是我陪着去的，就在陈老师的办公室。那天讨论的结果，应该说是坦率的，但是我认为谁都没有改变谁。结果是陈老师坚持要把这些

村子住民迁走，干干净净地体现这个遗址。此前 ICOMOS 的司库乔拉·索拉（Giora Solar）来西安的时候，我也就这个问题向他请教过。乔拉的观点其实很开放，认为村子对重要遗址有占压的，那是需要协商搬迁的；影响不大的，是可以共存的。这个观点我们也都转述给陈老师，但是说服不了。我觉得这个问题是国际古迹遗址保护领域比较前沿的一个问题，留什么，不留什么，怎么评价历史，历史的价值、遗产的价值到底以什么方式体现？我觉得都值得讨论。

采访人：肖老师，关于哈罗德，您记忆中有什么让您觉得比较印象深刻的事？

　　肖莉：那我就先说个故事吧。有次我和哈罗德两个人从北院门回来，在一块儿吃饭的时候，可能看见旁边有孩子和父亲在一起，就问哈罗德是不是国外的孩子对父亲都是直呼其名？他说女儿莉迪娅（Lydia Høyem Anker）小的时候也是一直管他

cess by many Chinese experts. What is your opinion on this dispute?

Liu Kecheng: Due to the difference in professional and academic backgrounds, there are different opinions as to what is authenticity and integrity. I had a lot of arguments with the archaeologists. Many Chinese archaeologists are fond of choosing one specific section in history, instead of seeing the whole history. Then the argument is: should we keep the entire history of the site or just focus on the authenticity and integrity of a certain historical period of the site? I think this issue is highly controversial, not that it has already become a common consensus in China. Even some prominent scholars are very much divided on this issue.

I had arranged for Harald to meet Director Chen Tongbing, who was in charge of the project of the Han Chang'an heri-

tage site. Before they met each other, I had explained Harald's opinion on the removal of the villages from the heritage site to Director Chen. I was there with Harald and Director Chen when they had the meeting. I would say they had quite open discussions about the issue, but still, neither side was convinced or changed their viewpoint. Director Chen still insisted on relocating the villages and presenting a purified heritage site. As to this issue, I also discussed it with Mr. Giora Solar from ICOMOS. He was pretty open about it. He said that all the villages may not need to be removed and that the villages which were compatible with the heritage site could stay. I also conveyed Mr. Giora Solar's opinion to Director Chen, but it didn't change her decision. I think these issues are some of the typical points of contention in the field of heritage site

叫爸爸的。然后我问他什么时候就不叫他爸爸了。他说，因为有次他要到中国来，莉迪娅不想让他离开，哭着闹着不让他走，但是他不得不走，自此以后她就直呼其名，再也不叫爸爸了。我说那一定是让莉迪娅伤心了。这件事让我感触很深——因为要来中国做事，让孩子抱怨，留下这样的一个记忆。反正挺感动我的。

我们在西安做历史街区和民居保护，是大学毕业（1984年）以后，当时做北院门、德福巷，做了好些，只是那个时期基本上是我们研究哪片区域，那片区域很快被开发了。本来是想保护，结果每一个都是这样的结果。当时刚好是西安旧城和低洼地区改造，很多开发商就想趁这个机会把很多老民居一块儿拆掉，然后我们就努力去测绘。那时候一周只有一个星期天，还不是后来的双休日，所以没多少时间，只能每个周日想办法去测绘。但是根本

来不及，很多很好的老四合院就这样被拆掉了。当时很愤怒，很生气，发誓说自己再也不要管西安城里的这些历史民居了，因为我们管不了。尤其是每当我们想去调研保护什么的时候，结果反而导致那些老房子更快速度地被拆掉，有一种引狼入室的负疚感，感觉还不如不干。

然后那时候哈罗德来做北院门的项目，遇到不少困难。我当时就跟刘老师说，看看能不能帮助他。那时确实只是觉得，如果能帮他做点儿事，让项目能够顺利进行，我就知足了，至于能不能保护好，我真的不知道。就是这样一种心态，我参与了北院门这个保护项目。

那时候，哈罗德好多次到西安来的时候，可能因为原来中方对接单位答应过什么，后来又不做了，很多事情推进得非常困难。有一天也是跟他一块儿吃饭的时候，哈罗德就很有感触地说：我这件事情是

conservation. What to keep and what not to keep? How to define the values of the site and history? In what way is the value of history and heritage reflected? I think it is worth discussing.

Interviewer: Professor Xiao, can you tell us something you remembered about Professor Harald? Something that impressed you most?

Xiao Li: I will start with a story that I remember. Once when Harald and I finished our work at Beiyuanmen and had some food at a restaurant, a man and his daughter had just sat next to us. The girl reminded me of a question, so I asked Harald if it was true that children in western countries could call their parents by their first names. Harald said that his daughter Lydia used to call him "Papa" when she was little. Then I asked Harald when Lydia stopped calling him that. Harald said that once he was to leave for

China, but Lydia did not want him to go and cried hard. Since then, Lydia called him by his first name and never called him "Papa" again. I said that the time must have hurt the little girl's heart. The story kind of touched me because it was a price Harald had to pay when heading to China to do something meaningful for him.

Since we graduated from college in 1984, we began the documentation work of historic districts and residential areas in Xi'an, including Beiyuenmen Street, Defu Alley, and so on. It was quite frustrating because most of the historic districts and residential areas we had researched were destroyed by real estate developers. At that time, it happened to be the renovation of the old city and terrain areas of Xi'an,

不是做错了？为什么我一个外国人拿着钱来帮你们保护这些民居，你们都这么不支持？是不是我做错了？我这么做有没有意义？那时候哈罗德也就60岁出头吧，作为建筑师属于正当年的时候。只是当时我觉得这个问题很难回答，因为我自己也仅仅是以一种帮忙的心态来的，我没法说其他人是怎么去想这件事儿。当看到哈罗德教授说自己是不是做错了，你们中国人是不是认为我特别傻，那一瞬间我特别感动。我当时就跟他说，我觉得你这件事儿真的是很有意义的。我们不做，因为我们觉得自己没办法推动这件事情，根本是螳臂当车。但也许你来做结果可能会不一样。从另一方面讲，我觉得有的人活了一辈子，仅仅是为自己活着；有些人拼命地工作、生活，是为了自己的家人生活得好一点；有的人做了很多工作，可能是为自己的国家；但确实有一种人，是真的为人类在做事情，那你就

是属于这种人，其实挺伟大的。我不知道哈罗德教授还记得不记得这件事。我觉得这也是让我觉得我必须得坚持，必须得把这件事情做下去的一个很重要的理由吧。

采访人：肖老师刚才的回忆很感人。我们还有一个问题：今天回想和哈罗德这么多年时光的合作的时候，有没有什么觉得比较遗憾的东西？

刘克成：我也是奔六的人了，今年58岁了。我觉得谈遗憾这个问题，可能跟二三十岁的时候想这个问题不太一样了。今天这个场合也回忆了很多东西，就是今天回想哈罗德当年跟我谈的一些话，当时的那个理解，其实跟今天的理解也不完全一样。

如果说遗憾，可能这个世界上任何一个有追求的人、执着的人，遗憾是一定有的，因为他不会那么容易满足。哈罗德就是一个对人对事特别真诚和认真的人。就我自己来说，我也是一个比较固执、认

and many real estate developers wanted to take advantage of this opportunity to demolish the old courtyards. We could only do the survey and documentation of the old buildings on Sunday because at that time people worked on Saturday as well, and we only had free on Sunday. Before we finished, many old courtyards had already been demolished. We felt so angry, desperate, and helpless. At the same time we felt guilty because whenever we tried to save one, it ended up being torn down. We kind of felt that it was we who attracted the real estate developers.

Harald had met with many difficulties as well during the Beiyuanmen Project. I talked about this to Kecheng and said that we shall try our best to help him. But frankly speaking, even though I wasn't sure if we could succeed in preserving the courtyards, we still joined the project.

Harald had been in China several times at the time, but things didn't go forward quite smoothly. Sometimes for some reason, the Chinese counterparts did not do the things they had promised. Once when I had dinner with Harald, he began to question himself. ¨Have I done something wrong? As a foreigner, I came to China with the money to help you to preserve those courtyards, but still, I didn't get support. What's the point of doing this?" I remember I didn't know how to answer those questions. I had joined the project to help, and it was not appropriate for me to give any comment. Harald was at that time only in his early sixties, sort of the best age as an architect. I felt very touched as well when I heard him saying that he came to China to help us but began to question himself. All I could say

真的人。很多事情成也罢，不成也罢，因为没有达到自己所有的期待，觉得遗憾这是正常的。但就我今天的理解来说，也许这就是生命和生活的常态。其实每个社会每个人都是这样过来的。能不能不要把遗憾当作抱怨，把这些东西化作一种力量，仍然能够平和地继续生活下去？对这个社会，对周边的人和事，还是要有一种乐观和正面的期待更重要。这是我从哈罗德和挪威同行那里学习到的一个很重要的事情。不管这个世界怎么样，还是要积极地面对，还是要乐观地看待，还是要尽自己的努力去做到最好。我觉得可能从这个角度来说，这是我今天的认识。所以说遗憾多少，说成功多少、失败多少，其实没有太大意义。

从另一个方面来说，我跟哈罗德所有的接触历程，其实也正反映的是中国跟挪威的那种差别。我们好像都是赶命似的，一件事接着一件事，一个快节奏的生活，快节奏的工作。而每次遇到哈罗德和在挪威的时候，我觉得他就好像一个刹车器，让我能够安静下来，想想到底什么才是正常，怎样才是一个真正有意义的生活和工作的方式。我觉得哈罗德于我，一方面像一个父亲；另一方面，就是他能够让我自己在今天这样的一个环境下保持一种平静从容。这是我跟哈罗德相处几十年学习到的特别重要的东西。

今天的重点，就我来说，是一个年轻一辈向一位老者致敬的谈话。所以我觉得更重要的是说说哈罗德先生他教给我、感染我的东西。原本计划要写点东西的时候，就是打算写点小事。这真的是我今天的一个看法：专业上的得失、对错问题其实可以交给时间，交给历史。而人与人、朋友之间，那些有关日常情感的小事才是最难忘的。

（全文完）

---

at that time to Harald was that what he did was very meaningful. I told Harald that we didn't have the confidence to do what he did just because we knew it would be like swimming against the tide. I said things could be different when a foreigner was involved, and maybe you could succeed because you are a foreigner. I sincerely think that Harald was doing something good for mankind and greatly admire him for that.

Interviewer: What Professor Xiao just told us touched us as well. Thinking back on the years you have been working with Harald, are there any regrets?

Liu Kecheng: I am 58 years old now. I think talking about regrets may not be the same as when I was twenty or thirty years old. The occasion today also recalls a lot of things, that is, today I think back to some of the words that Harald said to me back then, my understanding at that time, was not the same as it is now.

When pursuing goals, I think anyone would have regrets to some point since we are not easily satisfied. Harald is very sincere and hard-working. As I understand it today, it is normal to have some regrets when many things don't meet all your expectations and that may be a part of life. Are we able to not take the regrets as complaints, but turn them into strengths? It is more important to be positive about the people and things around us. This is a very important thing I have learned from Harald and other Norwegian friends. No matter what, we should face the world with a smile and do our best. In that sense, regrets are not so important.

Through my contact with Professor

Harald, I also see the difference between China and Norway. In China we have got used to the high-paced life, rushing and stressing from one thing to another. When I was with Harald, it felt like Harald was a car brake for me, asking me to slow down and truly enjoy my work and life. For me, Harald is not just my colleague, but a friend and a sort of father figure to me. I have learned from him how to slow down and follow my own pace.

The focus of our talk today, as far as I am concerned, is an architect of a younger generation paying homage to a senior one. I think it's more important for me to talk about what Harald taught me and how he influenced me. What I originally planned was to write about some small details that happened between us. That is my view of reflection today: the professional gains and losses, right and wrong, can be left to time and history. Only those small details of interactions between people and friends in our daily lives are the most memorable.

(End of the interview)

后记

通过本书的文字和图片，我试图说明，在至今已三十五载的漫长时光中，在这样一个变化迅疾剧烈的时代中，我是如何认识和面对中国社会的。作为一个外国人，在与自己的挪威文化背景截然不同的中国语境下，我如何置身其中。启发来自方方面面：了解我们彼此的文化有哪些异同，过去与现今的生活如何交织在一起，人们在日常生活中如何对待时间的流逝，普通人如何有尊严地料理自己的生活……我了解到很多中国的实际状况，但于我而言，也许更重要、更个人化的是自己的身份认同如何因面对"他者"而被影响，这正是在融入各类活动时与他人相遇的结果。

在首次抵达中国和在中国逗留期间，正如本书最初几章所提及的那样，我迫不及待地想了解作为一个外国人，与中国文化相遇会是怎样的情形。而今，在多年经历这种相遇的背景下，我对跨文化沟通和良好对话的前提有了如下思考：

在这么多年来思考这个问题之后，有三个现象显而易见。当不同的人相遇时，身份认同、尊严和尊重以及这三者之间的相互关系，似乎是人与人沟通的关键。而最重要的是随着时间推移，人在与他人的共存中找到自己的身份认同。身份认同感自觉或不自觉地发展变化，身份认同的形成，无时无刻不在发生，无处不在，人人不免。这可能是一个缓慢的过程，也可能很迅速。但我认为，利用身份认同发展潜能的先决条件是在自己的身份认同形成过程中找到支持这种感受的连续性，从而得以成长。这既是挑战，也是相互慷慨的馈赠，是人不断自问"我是谁？"而发展成为人的潜力。我，作为一个人，什么是永久的？什么是变化的？

当别人不仅仅因为你在社会中的地位，而将你作为一个个体的人去尊重，就会感受到一种基本的尊严。为了让我们感到自己扮演着一个有意义的角色，我们是社会上受人尊重的成员，这种尊严是最基本的。如果我们想去积极促成同胞的身份认同，那么彼此间的相互尊重必不可少。

我要强调的是，在中国这些年，我所受到的不变的热情款待至为重要。我主要指一直乐于支持我们活动的合作方、友人和同事。这种持久的热情好客给我在中国的所有时光留下愉快的印象，使我形成了我的自我认同，感受到了尊重，也获得了为人的尊严。

EPILOGUE

Through the text and the pictures in this book, I have tried to illustrate how it has been to meet and face Chinese society for a long period of time, almost 35 years now, a period which has been characterized by rapid, almost violent changes. How it has been to muddle about as a foreigner in the Chinese context, which is rather different from my own background, the Norwegian culture. It has been enlightening in various respects; learning what is similar and what is different in our cultures; how the life of the past and life of the present are interwoven; how people relate to the flow of time in their daily life; how common people manage their lives with dignity. I have, of course, learned much about the actual situation in China, but maybe more important for me, and more personal, is how my own identity has been influenced and colored by facing "the others," first and foremost as a result of meeting other people when being integrated with various kinds of activities.

During the arrival and the first stays in China, as mentioned in the introductory chapters, I was eager to learn how it would be, as a foreigner, to meet the Chinese culture. Now, on the background of the many years of experiencing this encounter, I will briefly reflect on communication and the premise for good dialogues across cultures.

Thinking of this issue in the aftermath of all these years, three phenomena have become evident. When different people meet, *identity, dignity*, and *respect* – and their reciprocal relationships – seem to be crucial in the communication between people. And most important: Over time one finds one's own identity in the coexistence with other people. The feeling of identity, consciously or not, changes by evolving. Identity formation takes place without stopping, everywhere and for everyone. It may be a slow process and it may be rapid. But a prerequisite for making use of the potential of identity development is, I believe, to find and support the feeling of continuity in one's own identity formation, and thereby grow. This is a challenge as well as a generous gift, a potential for developing as a human being, always asking "who am I?" What is permanent in me as a person, and what is changing?

A fundamental *dignity* is felt when other people show *respect* for you as an individual person; not just because of your position in society. In order to feel like we play a meaningful role and that we are respected members of society, our dignity is fundamental. Communicating mutual respect for each other's dignity

330

is essential if we seek to positively contribute to the identity formation of fellow beings.

I will underline the importance of the unwavering hospitality I have met during these years in China. I refer mainly to counterparts, friends, and colleagues who have always been willing to support our activities. This permanent hospitality has pleasantly impacted my impression of China during all my time there and made it possible to gain and develop identity, a feeling of respect, and dignity.

致谢

许多人员机构参与了本书所谈涉的活动。我不能尽述每一个人的名字，但他们从未被遗忘。

书中各项任务得以开展，有赖于中国众多友好机构和个人的支持。我想特别感谢以下诸位：中国城市规划设计研究院的王景慧总规划师、张兵所长，中国建筑设计研究院建筑历史研究所的陈同滨所长，清华大学的吕俊华教授、孙凤岐教授、张杰教授、王韬博士，东南大学建筑学院的董卫教授及其同事，西安建筑科技大学的校领导和张似赞教授、刘克成教授、肖莉教授、刘辉教授等众多师生，以及张西元等西安回民历史街区保护项目办公室的各位同人。特别感谢张少一为我们在西安驻留期间所有活动提供的帮助。总之，感谢所有参与合作项目的政府人员和专业人士，以及我们工作过的村庄和城市的当地居民，谢谢你们！

与西安建筑科技大学的定期联合教学及后续科研活动需要一个合作机制。多年来，挪威科技大学（NTNU）在西安建筑科技大学设有一个办公室，由一名秘书负责协调工作。在刘克成院长的慷慨支持下，西安建筑科技大学建筑学院为我们提供了一间可供长期使用的办公室。大学的青年建筑师兼任秘书，为我们解决各种实际问题，安排开展联合教学活动所需的各项工作。他们分别是许东明、王宇和苏静。这个分支机构在我们两所大学间建立了坚实的联系，并促成新的合作项目。

与法国同事的合作也是愉快的，其中包括巴黎建筑、城市与社会研究院（IPRAUS）的皮埃尔·克雷蒙（Pierre Clement）教授，以及波尔多国家建筑和景观高级学院（ENSAP）的法约勒·吕萨克（Bruno Fayolle Lussac）教授、让-保罗·卢贝斯（Jean-Paul Loubes）教授和人类学学者伊内卡·阿梅兹（Ineka Amez）。

挪威方面也提供了很多宝贵的帮助。特别感谢挪威开发合作署（NORAD）和挪威王国驻华大使馆；建筑师阿蒙·拉尔森（Amund Sinding Larsen）为西安鼓楼回民历史街区项目提供的帮助；我所在的挪威科技大学的领导、同事和学生，特别是我多年的好同事艾尔·格丽特莉（Eir Grytli）教授、达克·尼尔森（Dag Nilsen）教授、谢尔-哈瓦德·布莱顿（Kjell-Håvard Bråten）副教授和丽莎贝特·索里娅（Lisbet Sauarlia）副教授。

马丁·霍耶姆（Martin Høyem）负责了对本书英文原文的校订。作为值得信赖的朋友和专业人士，许东明将本书译成中文。王韬、董卫对中文译稿进行了细致校订，并给出他们的意见。丹尼尔·因凡特（Daniel Infante）对中文移译英文内容做了专业校对。王宇从中协调帮助，使本书最终得以顺利出版。

在本书最后完成过程中，我完全仰赖上述四位中国朋友的帮助和支持。我也特别感谢朝华出版社和责任编辑刘小磊先生为书稿出版付出的辛勤劳动。成都理工大学的张洁教授和成都信息工程大学的李伟彬教授为本书英文稿的校对出力甚殷。感谢群岛 Archipelago 负责人秦蕾女士和平面设计师马仕睿先生为本书排版设计所做的工作。

最后，感谢我的妻子玛丽娅·安珂（Marie Louise Anker）。这些年来，无论在生活还是在工作中，她一直是我的坚实后盾，也是我的专业伙伴。

ACKNOWLEDGEMENTS

Numerous persons and institutions have been involved in the activities described in the book. I choose not to name everyone, but while most of them are not mentioned, they are not forgotten.

Benevolent institutions and persons in China have made it possible to carry out all kinds of tasks. I would like to specifically mention: The good colleagues Wang Jinghui and Zhang Bing of China Academy of Urban Planning and Design, Director Chen Tongbin of China Architecture Design and Research Group, Professor Lyu Junhua, Professor Sun Fengqi, Professor Zhang Jie, and Dr. Wang Tao at Tsinghua University, Professor Dong Wei and his colleagues at Southeast University, Nanjing (SEU), Professor Zhang Sizan, Professor Liu Kecheng, Professor Xiao Li, Professor Liu Hui, and their staff and students, as well as University leaders of Xi'an University of Architecture and Technology XAUAT, the staff of the project office in the Drum Tower Muslim District (DTMD), headed by Zhang Xiyuan. In general, all the politicians and professionals involved, as well as residents of the villages and urban districts where we worked – a great thank you to you all. A special thanks to Zhang Shaoyi who was our loyal friend and helper through all our activities and stays in Xi'an.

The regular joint teaching with XAUAT and later research activities required an organized cooperation structure. For many years the Norwegian University of Science and Technology (NTNU), established a branch office at XAUAT, manned by a secretary who helped with logistic matters. XAUAT also offered a permanent office for our staff when we worked there, thanks to the generosity of Dean Liu Kecheng and Professor Xiao Li. Young architects from the school served as secretaries and made it possible to overcome practical problems as well as arrange what was necessary for carrying out our joint activities. I would like to mention the names of those helpful persons in succession: Xu Dongming, Wang Yu, Su Jing. This branch office created solid ties between our universities and facilitated new joint activities.

Cooperation with French colleagues have also been a pleasure, among them Professor Pierre Clement in IPRAUS, Paris, and Professor Bruno Fayolle Lussac, Professor Jean-Paul Loubes and anthropologist Ineka Amez of École Nationale Supérieure d'Architecture et de Paysage de Bordeaux.

Contributing with valuable help on the Norwegian side I will mention the Norwegian Agency for Development Cooperation (NORAD) and the Norwegian embassy in Beijing; architect Amund Sinding Larsen for his kind help to access the DTMD project; leaders, students and colleagues of my university, NTNU, and I would particularly like to mention my good colleagues through many years, Professor Eir Grytli, Associate Professor Dag Nilsen, Associate Professor Kjell-Håvard Braaten and Associate Professor Lisbet Sauarlia.

Martin Høyem has through proofreading been a great help to clarify language and thereby the content of the book. Dongming has translated the book to Chinese as a trustworthy friend and professional in both as to content and language translation. Wang Tao and Dong Wei have proofread the translation and given comments on the whole manuscript. Mr. Daniel Infante has been a steady proofreader where Chinese texts are translated into English. And Wang Yu coordinated and helped to make the book finally published.

During the finishing period of producing the book, I have been totally dependent on the help and support of four hardworking musketeers: Professor Dong Wei, Executive Editor-in-Chief Wang Tao, Senior Researcher Wang Yu, and Adjunct Research Professor Xu Dongming. My special thanks as well to Blossom Press for making it possible to publish this book. As the responsible editor, Mr. Liu Xiaolei has worked hard to improve the manuscript. Professor Zhang Jie from Chengdu University of Technology and Professor Li Weibin from Chengdu University of Information Technology took on the great labor of proofreading the entire English manuscript of this book. In the very last stage of the production, the layout work is performed by Archipelago Studio, mediated by the leader, Ms. Qin Lei, and performed by graphic designer, Mr. Ma Shirui.

Finally, my wife, Marie Louise Anker: She has, all these years, been my steady, warm-hearted, and patient support and professional fellow, in life and in the field.

致谢

ABBREVIATIONS

DTMD	The Drum Tower Muslim District
ICOMOS	International Council of Monuments and Sites
NORAD	The Norwegian Agency for Development Cooperation
NTNU	The Norwegian University of Science and Technology
SEU	Southeast University
SINTEF	The Foundation for Scientific and Industrial Research
UNESCO	The United Nations Educational, Scientific and Cultural Organization
WHL	World Heritage List, UNESCO
XAUAT	Xi'an University of Architecture and Technology

REFERENCES

- Barth, Fredrik. Andres liv og vårt eget. Oslo: Universitetsforlaget, 1991.
- Bråten, Kjell Håvard and Høyem, Harald. The Resource System Method. NTH, Trondheim, 1991.
- China ICOMOS. Principles for the Conservation of Heritage Sites in China. Beijing, 2000.
- Dong Wei. An Ethnic Housing in Transition. Chinese Muslim housing Architecture in the framework of resource management and identity of place. NTH, 1995.
- Finocchiaro, Luca; Haugen, Tore. Beyond Comfort. Energy retrofitting of a historic house in Hou Ji Village. NTNU, Trondheim, 2017.
- Gehl, Jan. Livet mellom husene. Arkitektens Forlag. København, 2003.
- Høyem Harald, Erring, Bjørn and Vinsrygg, Synnøve eds. The Horizontal Skyscraper. Tapir Academic Press, Trondheim, 2002.
- Høyem, Harald. Drum Tower Muslim District project report. NTNU. Trondheim 1994.
- Høyem, Harald. Research on future potential - Xiyangshi street No.4, Xi'an. Report. Xi'an, 1999.
- Høyem, Harald; Braaten, Kjell Håvard; Landsem, Inger Anne. Da Bao Ji Xiang quarter, Xi'an. Project report. NTNU, Trondheim, 1992.
- ICOMOS. The Nara document on authenticity, 1994.

- ICOMOS. International Charter for the Conservation and Restoration of Monuments and Sites (The Venice Charter), 1964.
- Jankowaiak, William R. Sex, Death and Hierarchy in a Chinese City – An Anthropological Account. New York: Columbia University Press, 1993.
- Jun Jing. "Feeding China's Little Emperors: Food, Children and Social Change," Stanford University Press, 2000.
- Larsen, Peter Bille ed. World Heritage and Human Rights – lessons from the Asia-Pacific and global arena. Routledge, London and New York, 2018.
- Lussac, Bruno Fayolle, Høyem, Harald and Clément, Pierre eds. Xi'an – an ancient city in a modern world. Evolution of urban form 1949-2000. Paris: Éditions Recherches, 2007.
- NTNU and SEU. Beyond Comfort: Energy retrofitting of a historic house in Hou Ji village. Trondheim, 2017.
- Shaanxi Provincial Muslim Architecture and Design and Planning Institute. The research on renovation and reconstruction in Muslin district of Xi'an. Xi'an, 1997.
- Shaanxi Provincial Institute for the Design and Conservation of Ancient Architecture. Drum Tower - Muslim District, Xi'an, China. Survey drawings - no 125, Hua Jue Alley.
- SINTEF report STF22 A03506, April 2003. Environmental and Resource Analysis of the Drum Tower District.
- Wang Jing Hui, Zhang Jie and Høyem, Harald eds. Permanence and Change – theory and practice of urban conservation in China and Europe. Shanghai: Tong Ji University Press, 2014.
- Wang Shu and Lu Wenyu, Amateur Architecture Studio, Hangzhou. Exhibition in Arc en rêve, Centre d'architecture, Bordeaux, 2018.
- Wang Tao. A Social Perspective on the Reformed Urban Housing Provision System in China. Three cases in Beijing, Xi'an, and Shenzhen. NTNU 2004.
- Wang Yi. The Impact of Tourism in the Historic Town of Fenghuang, China, and the Changes in Buildings and Inter-building Public Space. NTNU, 2019.
- Wang Yu. How Do We Rebuild a Disaster damaged Heritage Settlement: A study of the Post-Earthquake Reconstruction of the Village of Taoping. A Traditional Qiang Settlement in Sichuan China. NTNU, 2015.
- World Heritage Centre. Operational Guidelines for the Implementation of the World Heritage Convention. Paris 2012.

- XAUAT and NTNU - Involvement of local people in a development and conservation process of a historic site - Case Han Chang'an city. Project description. Xi'an, 2014.
- Xi'an Qujiang Daming Gong Site Area Protection and Reconstruction Office, Xi'an Bureau of Cultural Heritage. Competition material – International Conception Design – National Relics Park of Tang Daming Gong Site. Xi'an, 2007.
- Xu Dongming. A Multi-Perspective Observation of Site Museums: Case study of Archaeological Site Museums in China, with Norwegian Example as Reference. NTNU 2018.

著作权合同登记号 01-2023-0124

图书在版编目（CIP）数据

　　永恒与变迁：挪威建筑师镜头下的中国：1985—
2019：汉英对照 /（挪威）哈罗德·霍耶姆著；许东明
译.--北京：朝华出版社，2023.1
　　ISBN 978-7-5054-5126-1

　　Ⅰ.①永… Ⅱ.①哈… ②许… Ⅲ.①古建筑－保护
－研究－中国－1985-2019－汉、英 Ⅳ.① TU-87

　　中国版本图书馆 CIP 数据核字（2022）第254633号

永恒与变迁：
挪威建筑师镜头下的中国（1985—2019）
（汉英对照版）

作者　（挪威）哈罗德·霍耶姆
翻译　许东明
校译　王 韬 董 卫

选题策划　赵 倩
责任编辑　刘小磊
责任印制　陆竞赢 崔 航
设计排版　群岛 ARCHIPELAGO

出版发行　朝华出版社	开　本	710mm×1000mm　1/16
社　　址　北京市西城区百万庄大街24号	字　数	195千字
邮政编码　100037	印　张	21.25
出版合作　（010）68995532	版　次	2023年1月第1版
订购电话　（010）68996522		2023年1月第1次印刷
传　　真　（010）88415258（发行部）	装　别	平装
联系版权　zhbq@cicg.org.cn	书　号	ISBN 978-7-5054-5126-1
网　　址　http://zhcb.cipg.org.cn	定　价	98.00元
印　　刷　北京印刷集团有限责任公司		
经　　销　全国新华书店